This is a state-of-the-art collection of essays on the relation between probabilities, especially conditional probabilities, and conditionals. It provides new negative results that sharply limit the ways conditionals can be related to conditional probabilities. There are also positive ideas and results that will open up new areas of research.

The collection is intended to honor Ernest W. Adams, whose seminal work is largely responsible for creating this area of inquiry. In addition to describing, evaluating, and applying Adams's work, these contributions extend his ideas in directions he may or may not have anticipated, but that he certainly inspired.

This volume should be of interest to a wide range of philosophers of science, as well as to computer scientists and linguists.

T0245341

Probability and conditionals

Cambridge Studies in Probability, Induction, and Decision Theory

General editor: Brian Skyrms

Advisory editors: Ernest W. Adams, Ken Binmore, Jeremy Butterfield, Persi Diaconis, William L. Harper, John Harsanyi, Richard C. Jeffrey, Wolfgang Spohn, Patrick Suppes, Amos Tversky, Sandy Zabell

This new series is intended to be a forum for the most innovative and challenging work in the theory of rational decision. It focuses on contemporary developments at the interface between philosophy, psychology, economics, and statistics. The series addresses foundational theoretical issues, often quite technical ones, and therefore assumes a distinctly philosophical character.

Other titles in the series
Ellery Eells, *Probabilistic Causality*
Christina Bicchieri and Maria Luisa Dalla Chiara, *Knowledge, Belief, and Strategic Interaction*
Richard Jeffrey, *Probability and the Art of Judgement*
Robert Koons, *Paradoxes of Strategic Rationality*

Forthcoming
Christina Bicchieri, *Rationality and Coordination*
Clark Glymour and Kevin Kelly (eds.), *Logic, Confirmation, and Discovery*
Patrick Maher, *Betting on Theories*
J. Howard Sobel, *Taking Chances*
Patrick Suppes and Mario Zanotti, *Foundations of Probability with Applications*
Jan Von Plato, *Creating Modern Probability: Its Mathematics, Physics, and Philosophy in Historical Perspective*

Probability and conditionals

Belief revision and rational decision

Edited by

Ellery Eells
University of Wisconsin, Madison

Brian Skyrms
University of California, Irvine

CAMBRIDGE
UNIVERSITY PRESS

CAMBRIDGE UNIVERSITY PRESS
Cambridge, New York, Melbourne, Madrid, Cape Town, Singapore, São Paulo

Cambridge University Press
The Edinburgh Building, Cambridge CB2 8RU, UK

Published in the United States of America by Cambridge University Press, New York

www.cambridge.org
Information on this title: www.cambridge.org/9780521453592

First published 1994
This digitally printed version 2007

A catalogue record for this publication is available from the British Library

Library of Congress Cataloguing in Publication data

Probability and conditionals : belief revision and rational decision /
edited by Ellery Eells, Brian Skyrms.
p.   cm. – (Cambridge studies in probability, induction, and decision theory)
"Essays on probability and conditionals is intended to honor
Professor Ernest W. Adams" – Introd.
ISBN 0–521–45359–3
1. Probabilities.   2. Conditionals (Logic).   3. Belief and doubt.
4. Decision-making.   5. Adams, Ernest W. (Ernest Wilcox), 1926–   .
I. Eells, Ellery.   II. Skyrms, Brian.   III. Adams, Ernest W.
(Ernest Wilcox), 1926–   .   IV. Series.
BC141.P69   1994
160 – dc20   93-41163

ISBN 978-0-521-45359-2 hardback
ISBN 978-0-521-03933-8 paperback

Contents

Contributors

Ernest W. Adams is a professor of philosophy emeritus at the University of California, Berkeley.

Charles S. Chihara is a professor of philosophy at the University of California, Berkeley.

Ellery Eells is a professor of philosophy at the University of Wisconsin, Madison.

Alan Hájek is an assistant professor of philosophy at the California Institute of Technology.

Ned Hall is writing a dissertation in philosophy at Princeton University.

Richard Jeffrey is a professor of philosophy at Princeton University.

Vann McGee is a professor of philosophy at Rutgers University.

Judea Pearl is a professor of computer science at the University of California, Los Angeles.

Brian Skyrms is a professor of philosophy at the University of California, Irvine.

Robert Stalnaker is a professor of philosophy at the Massachusetts Institute of Technology.

Patrick Suppes is a professor of philosophy emeritus at Stanford University.

Ernest W. Adams. (Photo: Bill Adams, Albuquerque, New Mexico)

Introduction

This collection of essays on probability and conditionals is intended to honor Professor Ernest W. Adams, whose seminal work is largely responsible for creating this field of inquiry. More than one thinker has been struck by the idea that conditional probabilities provide a key for understanding the special role that conditionals play in language and thought. In this respect, we should mention Frank Ramsey, Bruno de Finetti, Hans Reichenbach, Richard Jeffrey, Brian Ellis, and Charles Stevenson. But it is Adams who has done the most to give substance to this idea – by sustained research into the logic of conditionals over a period of thirty years. This volume is a testimonial to, and continuation of, Adams' contributions to philosophy. The following essays not only describe Adams' work but also evaluate it, apply it in various ways, and in addition take his ideas in directions that Adams may or may not have anticipated, but that he nevertheless inspired.

The volume begins with an essay by Patrick Suppes on Adams' interpretation of conditionals. This piece puts Adams' work in perspective by considering various contexts in which interpretations of conditionals may have to be different: mathematics, science, ordinary language, and ordinary inference. Suppes here describes various contexts in which the question of the interpretation of conditionals is relevant and raises the issue of the generality of the Adams conditional.

Next, Brian Skyrms distinguishes two versions of the theory, both of which are parts of Adams' thought: The difference lies in whether one thinks of the probabilities involved as (subjective) degrees of belief or as objective chances. Skyrms takes these to have different intellectual functions, the former being a theory of updating conditionals, and the latter being a theory of decision-making conditionals. For the objective chance conditional, Skyrms shows how Adams' theory can be extended by construing conditional chance as a random variable and shows how the selection

1

function theory of Stalnaker is equivalent to this extended theory under the assumption of determinism.

The third contribution to the volume is a letter from Robert Stalnaker to Brian Skyrms, dated April 3, 1984. In this letter, Stalnaker generalizes the connection between the conditional chance account and Stalnaker selection functions (which holds in the case of determinism) to a connection between conditional chance and the set selection functions of David Lewis and others (which holds more generally). He also presents an example that is problematic on all these accounts and invites further generalization of the theory.

Next, Stalnaker and Jeffrey, in "Conditionals as Random Variables," advance an understanding of "matter-of-fact" conditionals that escapes the difficulties of "triviality results" but nevertheless gives a full theory of iterations and Boolean compounds of conditionals. The key idea is a certain construal of the values of conditionals as random variables. This random variable account is intended for the degree-of-belief conditional and is completely different from the random variable account of conditional chance that Skyrms proposes for the decision-making conditional. This is an intricate, ambitious, and promising development that makes contact with theories of Vann McGee and Bas van Fraassen.

In the fifth essay, Judea Pearl sheds further light on the probabilistic construal of conditionals. Especially interesting here is a proposal concerning how, based on Adams' thesis about the relation between probabilities of conditionals and conditional probabilities, a unified account of indicative and subjunctive conditionals can be developed. In addition, Pearl applies these ideas to the related issues of causation, decision, inference, and belief, and he elucidates the connections with his own important work on Bayesian networks.

The next three contributions to the volume establish senses in which the theory dubbed the "conditional construal of conditional probability" (CCCP) cannot be correct. As the CCCP sounds suspiciously like the leading idea behind the positive results in the preceding chapters of this volume, readers will be interested in reading those essays with an eye to the results in this section, and conversely. In their joint paper, Alan Hájek and Ned Hall first clarify several versions of the CCCP hypothesis and describe its intuitive pull. They then discuss recent "sources of suspicion," describing, reviewing, and categorizing some well-known and some more recent (indeed, some of their own) "triviality results" concerning CCCP. This is a valuable overview and review of the "triviality" issue involved in the CCCP hypothesis.

Hájek's "Triviality on the Cheap?" – the seventh chapter in this volume – ingeniously detects and clearly describes a general philosophical methodol-

2

ogy or recipe behind triviality results against the CCCP; the essay also includes a survey and appraisal of various such results. Next, Hall's "Back in the CCCP" offers yet three more interesting triviality results that shed light both on previous such results and on the kind of interpretation required on the conditional if CCCP is true. We think this triplet of essays represents the state of the art of triviality results for the CCCP.

In the ninth chapter, "The Howson-Urbach Proofs of Bayesian Principles," Charles Chihara presents a critique of the arguments used by two prominent Bayesians to defend fundamental Bayesian principles – including the use of conditional probabilities for updating by conditioning on the evidence. This essay raises questions relevant to the pragmatics of Adams' conditionals.

The tenth essay, by Vann McGee, asks what we do when we learn something that has no finite positive probability. To answer this question, McGee introduces infinitesimal probabilities via the nonstandard analysis of Abraham Robinson. These allow an elegant extension of Adams' framework for evaluating the "validity" of arguments containing conditionals.

Adams has made valuable contributions in other areas of philosophy that are not within the focus of this volume. These include the philosophical foundations of physics, decision theory and the foundations of subjective probability theory, the theory of measurement, and the foundations of geometry. The final essay in this volume is "A Brief Survey of Adams' Contributions to Philosophy," in which Patrick Suppes gives an overview of this work. Suppes was Adams' Ph.D. thesis adviser, and Adams was Suppes' first Ph.D. student. The volume concludes with a list of Adams' publications. This list is current only as of the beginning of 1993. Ernest W. Adams is currently pursuing vigorous research programs on many fronts, and we look forward to the fruits of these investigations.

Ellery Eells
Brian Skyrms

1

Some questions about Adams' conditionals

Patrick Suppes

I have liked, since it was first published, Ernest Adams' book on conditionals (Adams, 1975). There is much about his probabilistic approach that is of permanent value. Whatever modifications will be made – and certainly they will be for such an important concept – much will remain of his original ideas and analysis. The present paper is an amplification and extension of comments I made to Ernest Adams' own paper given at the 1987 meetings of the Pacific Division of the American Philosophical Association.

I have organized my questions under three headings: varieties of conditionals, problems of computation, and questions of psychological reality.

1. Varieties of conditionals

From Adams' various writings on probabilistic conditionals, it is not clear how he wants to treat the widespread use of conditionals in mathematical texts and especially in scientific texts. It seems apparent that he might not want to argue that books or articles on pure mathematics should use a conditional probability concept as the appropriate interpretation of conditionals, but the problem is certainly more complex in the case of scientific texts, especially those reporting the design and analysis of experiments. One can, of course, make the move of using probability 1 to a fare-thee-well, but that often seems awkward in cases of mathematics. In any case, the real point of my question is not this particular instance, which in fact Adams may have analyzed clearly in some place I have forgotten, but to

It is a pleasure to contribute to this *Festschrift* for Ernie Adams, who was my first Ph.D. student. I have learned a great deal from him over the years about a variety of topics, ranging from rational mechanics to the measurement of utility. I have especially benefited from the natural skepticism with which he greets half-baked philosophical ideas proposed by others, including myself.

raise the question whether or not a probabilistic interpretation of conditionals is going to work out well in almost all cases.

In this frame of mind, consider the following two sentences:

> *If Jones gets an ace, then he will get a face card on the next draw.*
> *If Jones gets an ace, then it is likely that he will get a face card on the next draw.*

It is not clear to me how, from a probabilistic interpretation of conditionals, one has anything like a clear qualitative algorithm for distinguishing between these two sample sentences. What is the impact on the probability of the main conditional to add to the consequent the standard phrase "it is likely that"? It would seem that there is a natural "computation" to move from the first sentence to the second, based on the way we think of "likely" as expressing a probability judgment. In a similar way, we can ring the changes on elaboration of the first phrase: *It is very likely, it is not likely, it is not very likely.* It is not that I think the handling of such phrases presents a problem, in principle, for Adams' conditionals; it is just that I am not clear how we are to think about such extensions, which are certainly needed to give an adequate account of the ordinary use of conditionals.

In a similar vein, I am not clear how to think about imperatives. For example consider this imperative: *If you are cold, close the door!* Similar problems are presented by any standard computer instruction. For example: *If register 6 is empty, jump to exit!* How are imperatives meant to work in relation to indicative conditionals? Still other cases of analysis are required when probabilistic considerations are directly embedded in the content of the antecedent of an imperative:

> *If it is likely to rain, close the window!*
> *If you think it is likely to rain, close the window!*

Again, I am just not sure how all these probabilities, easily embedded in the content of the antecedent or consequent of a conditional, are to interact in a systematic way with the probability of the conditional.

The general way of putting my question about the varieties of conditionals is this: What is Adams' characterization of the range of his conditional probability thesis for ordinary conditionals?

2. PROBLEMS OF COMPUTATION

Perhaps my most difficult question, at least in my own conception of Adams' conditionals, concerns problems of computation. For example, we ordinarily think of conditionals as being transitive, following the Law of Hypothetical Syllogism, but in general this is going to be true for Adams' conditionals only when the probability is 1. But as in the case of transitivity of causes, it is a useful and important topic to investigate when transitivity holds. My question is, What kind of results should be expected here? For a discussion of transitivity of non-Markovian causes, see Suppes (1986).

Of more concern to me than the question of transitivity is the general problem of how to compute. Simple gambling examples are too easy. Is the thesis that there are implicit upper and lower bounds on a collection of heterogeneous conditionals in terms of the conditional probability we can assign to the conclusion, or else how are we to think about the actual results expected from such inferences? Bayesians might offer an answer by sketching how they would put the various probabilities together, but it does not seem to me that Adams would necessarily want to adopt that answer. (I shall return to this point later.) Another possibility that might well be appropriate in his framework is to use a qualitative theory of probability that can, of course, nicely express the qualitative conditional probabilities. Of course, the problem here is that, in general, we do not have good rules for combining qualitative conditional probabilities when we put together an argument or a piece of discourse. To emphasize the fact that we do not is another way of saying how subtle and complicated it is to compute the probability of a conclusion coming out of a piece of discourse. My skepticism is not meant to imply that I think the usual interpretation of conditionals as material conditionals will work. This is an obvious shortcoming of standard logic, which has too little to do with most reasoning that takes place in ordinary talk. I do not mean to suggest by the questions or problems that I raise here that alternative theories are better than Adams', at least not the alternative theories now available to us, but I do want to suggest that there is a problem that needs solving in any theory of conditionals that claims to give a successful account of ordinary reasoning.

There is a reason for being doubtful that the exact numerical values of probability will matter in most ordinary informal reasoning, but then there is also the problem that the theory of qualitative probability is not well set up for combining inferences, except for the Bayesian, who can perhaps do it by brute intuition. Of course, even though I am a sometime Bayesian myself, it is a paradox of Bayesian thinking that it is only in cases of relative frequency data, resulting from nicely designed experiments, that we all move in the same way from a given prior to a good posterior. The reason is simple. Talk about being Bayesian really provides no explicit methods for evaluating conditionals of any complexity. A standard answer is that the Bayesians can do it because they have a joint probability, and all they have to do is conditionalize in terms of the joint probability distribution. This is, however, a piece of fantasy. We simply do not, in ordinary circumstances, carry around any serious knowledge of a joint probability distribution of all the great variety of events, states, or facts that are referred to in ordinary discourse. Bayesians do not really have a prior that encompasses all these matters; they just have pieces of a prior distribution – a marginal here, a marginal there – but certainly no complicated precise joint distribution.

Another way of putting my question, which is central in many ways to the viability of Adams' conditionals, is one that is very much in the spirit of the question that Adams himself addresses to traditional logic: Can the theory being given escape from the scholastic box into the real world? Much more complicated and realistic examples must be given over a variety of domains before we can wholeheartedly accept an affirmative answer.

As a way of making this general remark more particular, I cannot but be skeptical that Adams' intriguing conditional deficit formula can possibly be computed even approximately in connection with conditionals used in ordinary discourse. Let us remember that the conditional deficit formula is the difference between the probability of the disjunction $\neg A \lor B$ and the probability of the conditional $A \to B$, which can be translated into the deficit $(1 - P(\neg A \lor B)) \cdot P(\neg A)/P(A)$.

3. PROBLEMS OF PSYCHOLOGICAL REALITY

There is a strong tendency in the current theories of language production and comprehension to postulate that many matters are modular, although exactly how these modules are themselves defined or put together is still a matter of much speculation that obviously will require much further work, both theoretical and empirical. However, in examining a new theory of conditionals, it is natural to ask what kinds of problems would arise in processing, let us say, the production of conditionals if Adams' conditionals were most commonly used. At a glance, this would seem to be an easy question for Bayesians. It would just be the imposition of a joint probability distribution properly conditionalized in the final evaluation of the conditional probability of the uttered conditional. For the reasons stated earlier when the conditional was at all complicated, this story seems to have the elements of a wonderful fairy tale. First we form the complicated joint distribution, and then we go through the complicated process of conditionalizing. Moreover, it would be natural to think of indicating in prosodic features, at least in the case of spoken conditionals, the degree of confidence or belief one had in the conditional, as expressed in the underlying conditional probability. Does, however, such a connection between prosodic feature as a component of a speech act and the probability

8

assigned to a conditional have any systematic relation that can be observed? Perhaps the right way to think about Adams' conditionals as reflected in this kind of problem is that most of the conditional probabilities are close to 1, and therefore a kind of rough-and-ready computation is not too difficult.

But it seems to me that it is just in ordinary talk that we utter a lot of conditionals that we do not have great confidence in – to use another way of talking about their probable truth. We meander around in our conversations in all sorts of ways, expressing casual opinions and ill-thought-out ideas, and sometimes nothing much at all is to be found in what is said except the mere pleasure of talking. I give a number of examples of this last phenomenon in the section entitled "Grody to the Max" in Suppes (1984, pp. 170–2).

In these common cases perhaps it is right to think in terms of conditional probabilities that are far from 1. But then we have the problem of how seriously we should take them, and whether a numerical view of the probabilities is an appropriate one. It seems to me there is again a thicket of problems that are in all likelihood not insurmountable, but it is certainly not clear how to think through a more detailed formulation of the psychological theory of processing governing Adams' conditionals.

Perhaps the deepest problem of psychological processing for me is that it seems reasonable to suppose in the case of very many of the conditionals we utter that the evidence on which the conditionals are based is only partly summarized in the statement itself. In most practical situations it is too difficult to summarize with great care the complex evidence on which we base an assertion. If we make the assertion in conditional form, then we select certain items of information, but not simply because of their high probability or because they render a high probability for the conclusion, but for many kinds of reasons – ranging from the desire to persuade to the desire to use a popular argument in which we believe but which we think to be not as powerful as the real argument. Moreover, we have no rational scheme for selecting from our complicated network of beliefs exactly those that should be expressed verbally in forming a conditional. Further, as the dynamics of any situation changes, we also do not have rationally formulated mechanisms of attention telling us what we should watch for and what we should record in our network of beliefs. Of course there are obvious things we all notice because of their great saliency, but in ordinary situations, we find that our past history, our current preoccupations, and God knows what else play significant roles in that to which we attend. The conditionals we utter are reflections of such attendance, but no accurate mirror.

When, as in Adams' 1986 article, we restrict ourselves to high probability, there is another point that puzzles me, a point I made earlier in another context (Suppes, 1966). Consider the following two inference forms involving high probability, the first for material implication and the second for conditional probability:

$$\frac{\begin{array}{c} P(A \rightarrow B) \geqslant 1 - \varepsilon \\ P(A) \geqslant 1 - \varepsilon \end{array}}{P(B) \geqslant 1 - 2\varepsilon}$$

$$\frac{\begin{array}{c} P(B|A) \geqslant 1 - \varepsilon \\ P(A) \geqslant 1 - \varepsilon \end{array}}{P(B) \geqslant (1 - \varepsilon)^2}$$

For very small ε, $1 - 2\varepsilon$ and $(1 - \varepsilon)^2$ are quite close together. Surely such a small difference in itself cannot have much psychological significance. For Bayesians who assign at best a high probability slightly less than 1 to nearly any statement, the thesis about the probabilistic interpretation of conditionals is easily accepted, but the difference between conditional probability and material implication, as shown earlier, is of little importance. Given Adams' careful and, to my mind, devastating arguments against many ideas put forward in support of counterfactuals, I feel rather confident that he has sound objections to the too-simple analysis I have just given.

4. CONCLUDING REMARKS

I want to emphasize as strongly as I can that the many skeptical questions that I raised about the current status of the theory of Adams' conditionals is not meant to express any strong preference for another theory, because I think that Adams is off to an excellent start and has, moreover, backed it up with a great deal of systematic and technical analysis – see Adams (1986) for an example of the way he has successfully worked out details of the theory, and also the references there to earlier work.

My point is to stress how far I think we are from anything like an adequate theory of how conditionals are used in ordinary talk. There may be some thin version of rationality that would let Adams' theory of conditionals with high probability, as developed in the 1986 article, be seen as adequate, but I am an unreconstructed advocate of a much thicker concept of rationality – in spite of many things I have written myself in the past. We are, in my judgment, still very far from a satisfactory theory of rational talking, let alone rational decision making, once we take into account detailed and realistic concepts of how our attention mechanisms work and how we process information and for what purposes. In the meantime, I hope that Ernest Adams will continue to march forward and lead the way for the rest of us to follow in the use of probability in developing a more realistic theory of ordinary language.

REFERENCES

Adams, Ernest W. (1975) *The Logic of Conditionals.* D. Reidel, Dordrecht.
Adams, Ernest W. (1986) "On the Logic of High Probability." *Journal of Philosophical Logic* 15: 255–79.
Suppes, Patrick (1966) "Probabilistic Evidence and the Concept of Total Evidence." Pp. 49–65 in *Aspects of Inductive Logic*, ed. J. Hintikka & P. Suppes. North Holland, Amsterdam.
Suppes, Patrick (1984) *Probabilistic Metaphysics.* Blackwell, Oxford.
Suppes, Patrick (1986) "Non-Markovian Causality in the Social Sciences with Some Theorems on Transitivity." *Synthese* 68: 129–40.

2

Adams conditionals

BRIAN SKYRMS

1. ADAMS' TWO THEORIES OF CONDITIONALS

Ernest Adams has developed a theory in which conditionals do not function as factual claims but rather as bearers of conditional probability. This is a *pragmatic* theory rather than a semantical theory. The pragmatic framework is precise and probabilistic and quite different from what one usually finds discussed under the name "pragmatics" in contemporary linguistics and the philosophy of language. I believe that Adams' ideas are ahead of his time and that the contemporary philosophy of language still needs to catch up with them.

There are really two theories, depending on whether one interprets the probabilities in question as degrees of belief or as objective probabilities (propensities, or hypothetical limiting relative frequencies). Commentators often have taken him to be advancing the first sort of theory, whereas in fact he has always had an interest in the second. I believe that both theories are right, but they are right for different purposes.

According to the first theory, conditionals carry probability values relevant to *updating* our degrees of belief. In the most basic case we update by applying the rule of conditioning. On observing that some evidential proposition[1] e is true, we take as our updated probability of an arbitrary proposition q our old probability of q conditional on e. Application of the rule in statistical inference has almost the status of a conditioned reflex. On the epistemic version of Adams' theory, the relevant value – $\mathrm{pr}(q|e)$ – is just the value we carry around as our assessment of the credibility or assertability of the conditional: "If e, then p."

Where probabilities are interpreted as objective probabilities, the probability conditionals from act to consequence are central to the theory of rational decision (Adams, 1988). The conditional probability $\mathrm{pr}(C|A)$ is the conditional chance of consequence C on act A, and an act A can be thought of as tantamount to choosing an objective lottery over consequences

13

with chances $\text{pr}(\cdot|A)$. One wants to choose the act that will maximize the objective expected payoff of the associated lottery:

$$U(A) = \sum_{i=1}^{n} \text{pr}(C_i|A)U(A \wedge C_i)$$

Objective expected payoff is thus a species of expected utility, and, as Adams (1988) notes, an appropriate formal system is that of Jeffrey (1965), with the probabilities suitably reinterpreted. Calculating the objective expected payoff for act A consists in contemplating for each possible consequence the value of the associated objective probability conditional – "If I were to do A, consequence C would ensue" – and using those values to weight the payoffs that attach to the consequences. This is both a natural way and a correct way in which to evaluate options when the conditional chances are known.

Thus, Adams' two theories display how conditionals are central to two key aspects of our intellectual life: learning from experience, and deliberating about rational choice. Probability conditionals are so useful in these contexts that if we did not already have them in our language, we would have to invent them.[2]

2. DECISION THEORY AND VALUES OF DECISION-MAKING CONDITIONALS

The foregoing account of objective probability conditionals and rational decision applies when the conditional chances are known, but what if the decision maker is uncertain about the true conditional chances? We can pursue an answer by extrapolation of Adams' ideas.[3] Satisfactory treatment of this case requires an additional layer of complexity. The decision maker can do no better than evaluate according to her degree-of-belief expectation of objective expected value.[4]

Let us suppose that the decision maker has m hypotheses H_j about ways in which the objective conditional chances could be fixed. Let us suppose that the act together with the consequence determines the payoff. Then the measure of merit for an act is

$$U(A) = \sum_j \text{DB}(H_j) \sum_i \text{pr}_j(C_i|A)U(A \wedge C_i)$$

where DB is the degree of belief and pr_j is the conditional chance according to hypothesis H_j. This is a degree-of-belief expectation over an objective chance expectation. It is one general version of causal decision theory (Gibbard and Harper, 1978; Lewis, 1980; Skyrms, 1980a).

According to this general theory of rational decision, what is the

14

appropriate value for the decision-making conditional in the general case in which conditional chances are unknown? That is to say, what is the value Val(if A, then C) such that

$$U(A) = \sum_i \text{Val(if } A, \text{ then } C_i)U(A \wedge C_i)$$

The answer is that

$$\text{Val(if } A, \text{ then } C) = \sum_j \text{DB}(H_j)\text{pr}_j(C|A)$$

Val must be the subjective expectation of conditional chance. Then

$$\sum_i \text{Val(if } A, \text{ then } C_i)U(A \wedge C_i)$$

$$= \sum_i \sum_j \text{DB}(H_j)\text{pr}_j(C_i|A)U(A \wedge C_i)$$

$$= \sum_j \text{DB}(H_j)\sum_i \text{pr}_j(C_i|A)U(A \wedge C_i) = U(A)$$

as desired.

This account smoothly extends Adams' decision-making conditional of Section 1 to contexts where conditional chances are unknown. The decision-making conditional carries a value that is not a probability of truth. The computation of that value here is slightly more complex, but its role in deliberation is the same as before. It is the value appropriate to weighting the possible payoffs in assessing the utility of an act.[5]

3. CONDITIONAL CHANCES AS RANDOM VARIABLES

What is chance? Whatever one's theory, chances are determined by something about the world external to the decision maker. The determinants of optimal decision-making chances partition the possible worlds, with the chances being the same in each cell of the partition. For simplicity here, we assume that the partitions are finite. We can model the decision maker's uncertainty about the true chances by construing the "chances" as degrees of belief conditional on some partition,[6] and the degrees of belief about chances are degrees of belief about the true member of the partition.

Let us suppose that we have the appropriate partition for determining the chances of the outcomes of interest.[7] What then is conditional chance – in particular the chance of consequence conditional on act? Here we want the antecedent of the conditional – the act – together with the background conditions jointly to determine the chances of the possible consequences. That is to say, we want a partition of background factors that is coarser

15

	Urn 1 has 6 Red and 4 Black Urn 2 has 1 Red and 9 Black	...
Urn 1	Ch(R) = .6, Ch(R|Urn 1) = .6	...
Urn 2	Ch(R) = .1, Ch(R|Urn 1) = .6	...

Figure 2.1. Background partition.

than the chance partition, such that the common refinement of it and the act partition is sufficiently fine to determine those chances.

For a canonical example, let us suppose that a ball is to be drawn at random from an urn, with different payoffs attaching to balls of different colors. The decision maker chooses the urn from among a number of available alternatives. Thus the composition of the urns, together with the choice of urn, determines the chance of drawing a ball of a given color. The decision maker may be uncertain about the composition of the urns. She partitions her belief space according to the composition of the urns. This is the background partition. Within each of the cells of this partition, the conditional chance of a red ball conditional on it being drawn from urn 1 is just the proportion of red balls in urn 1. This is illustrated in Figure 2.1.

Some aspects of this treatment of conditional chance deserve special emphasis. One is that for the decision maker, conditional chance is *objective*. Conditional chances have a definite value at each point in the decision maker's probability space. *Conditional chances are random variables.* This is connected with the counterfactual aspect of the definition. Chances conditional on the ball being drawn from urn 1 are still defined in possible situations in which the ball is not drawn from urn 1. For example, suppose that the proportion of red balls in urn 1 is .6 in cell 1. Then the conditional chance of red conditional on urn 1 is .6 not only at those points in cell 1 where the ball is drawn from urn 1 but also at those points in cell 1 where the ball is drawn from a different urn.

Consider two special cases of this model. At one extreme, the decision maker knows which member of the background partition she is in. Then she knows the relevant conditional chances and is in the decision-making situation discussed in Section 1. She calculates expected utility as in Jeffrey (1965). At the other extreme, chance is replaced by determinism – the urns all contain balls of just one color. Then the relevant conditional chances are all 0 or 1, and all uncertainty is uncertainty about the background state. This is the sort of decision-making situation modeled by Savage (1954).

We can view the deterministic model in a slightly different light. Let us

16

call the points w in our probability space "possible worlds" and assign a distance d as follows:

1. $d(w, w) = 0$
2. If $w \neq x$ and w, x are in the same cell of the background partition, then $d(w, x) = 1$.
3. If w, x are in different cells of the background partition, then $d(w, x) = 2$.

Then d is a metric on the space of possible worlds, and the conditional chance of outcome on act in a world is equal to the truth value of the associated outcome proposition in the closest world in which the act proposition is true. For the deterministic case, metric d, and these decision-making conditionals, the conditional chance is equal to the truth value of the associated counterfactual conditional, as in Stalnaker (1968), and the decision maker's degree of belief in the Stalnaker conditional is equal to her "value" (subjective expectation of conditional chance) for the Adams-style decision-making conditional. In this guise, the deterministic case corresponds to the decision-making model proposed in Stalnaker (1981) and studied in Gibbard and Harper (1978).

4. FROM DECISION-MAKING CONDITIONALS TO NATURAL BAYESIAN CONDITIONALS

In the foregoing, the decision maker's options imposed a natural partition of the decision maker's probability space. We then proceeded to supply a partition of background factors such that background factors together with acts determined the chances of outcomes. This construction can be carried out just as easily when the members of the partition associated with the antecedent of the conditional are not options for the decision maker. We might have a *list of factors* that are jointly taken to determine the chances of an outcome: for example, a list of risk factors for suffering a heart attack in the next year. Let us suppose for simplicity that each factor is a variable taking a finite number of mutually exclusive and jointly exhaustive values and that the factors are logically independent. A *natural condition* will be the specification of a value for each member of some subset of the factors. Then for a natural condition to evaluate the conditional

If Condition, then Outcome

we would naturally take the appropriate background partition to be the partition generated by all the remaining factors that are not specified in the condition. Then the account of the *Value* of the conditional proceeds just as in Section 3.

Because the antecedents here may vary in specificity, we need not just one background partition but rather a *natural family of partitions*. As more information is hypothesized in the antecedent, less need be filled in by the background. Formally, the family of partitions is a function that maps every natural condition onto the partition whose cells represent joint assignments of values to the factors not specified in the condition. The fact that the background is merely "taking up the slack" in the specification of chance corresponds to a *monotonicity* property of the natural family of partitions so generated by a list of factors: If p, q are natural conditions and p entails q, then the background partition for q is a refinement of the background partition for p.

Again, it is worth looking at the special case where chance gives way to determinism. Here, as far as we are concerned, a possible situation can be taken to be individuated by the factors in our list together with the outcome. Then the case of determinism is the case in which the partition determined by all the factors has just one possible situation in each cell. The outcome in that situation is the outcome determined by the joint specification of all the factors corresponding to the cell. The natural family of partitions is then deterministic in the following sense: For every natural condition C, each member b of the background partition that the family assigns to C is such that the intersection $b \cap C$ contains just one point. The conditional chance of the outcome O is then either 1 throughout b or 0 throughout b according to whether the point in $b \cap C$ is in O or not.

Such a deterministic natural family of partitions defines a partial selection function. It is partial in that it is defined only for natural conditions. It maps a natural condition C and a situation w to a situation w' according to the following procedure: (1) Find the background partition that the family assigns to C. (2) Find the member b of this partition that contains w. (3) If w is in C, then $w' = w$; if w is not in C, then w' is the unique situation that is in $b \cap C$.

5. TOTAL BAYESIAN CONDITIONALS

Can we extend the foregoing theory from the case of natural conditions to that of arbitrary consistent conditions? There is such a general subjective Bayesian theory (Skyrms, 1984) that builds on the ideas of Adams (1975, 1976). It is assumed that we have a total family of partitions, indexed by the consistent conditions, and that for each consistent condition C and each cell b of the background partition assigned to C, $b \cap C$ has positive probability. It is no longer assumed that the partitions are generated by a list of natural factors or that the conditions are conjunctions of specifications of natural factors. We no longer have monotonicity of the

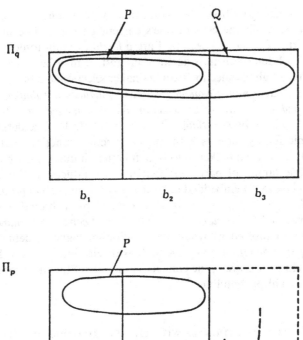

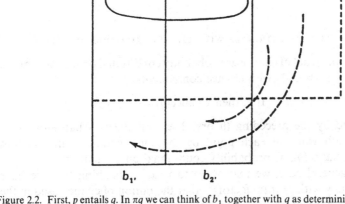

Figure 2.2. First, p entails q. In πq we can think of b_1 together with q as determining unconditional chance. *A fortiori*, b_1 together with p determines unconditional chance; πq will not serve as an appropriate partition for antecedent p because b_3 is not compatible with p. A natural way to move to a partition appropriate to p is to weaken background conditions b_1 and/or b_2 to $b_{1'}$ and $b_{2'}$, so that these now cover all cases originally in b_3.

family of partitions, but a weaker property that I have called *omonotonicity* is of interest. Let π be a family of partitions, and let $\pi(p)$ be the partition that the family assigns to condition p. Then π is omonotonic if whenever p entails q, every member of $\pi(q)$ whose intersection with p is nonempty is a subset of some member of $\pi(p)$. This is illustrated in Figure 2.2. Omonotonicity can be thought of as a kind of conservative principle for

19

moving to a new partition when confronted with a condition that is not compatible with all the cells of one's current partition. The monotonic natural partial families of Section 4 are *a fortiori* omonotonic.

The case of determinism is again of interest. It may be something of a surprise that a tight connection with Stalnaker selection functions continues to hold. *Every deterministic omonotonic family induces a Stalnaker selection function, and every Stalnaker selection function induces a deterministic omonotonic family of partitions* (Skyrms, 1984). Indeed, a deterministic omonotonic family induces a Stalnaker selection function that in turn induces it, and a Stalnaker selection function induces an omonotonic deterministic family of partitions that in turn induces it. For a simple conditional (with no embedded conditionals), the subjective *probability of truth of the conditional*, where truth is defined by a Stalnaker selection function, is equal to the *subjective expectation of conditional chance* of the consequent on antecedent, where the conditional chance is defined by the family of partitions induced by the Stalnaker selection function. This pins down the elusive nature of the connection between Stalnaker conditionals and conditional probabilities.[8]

6. CONDITIONALS WITH CHANCE CONSEQUENTS

Given the construal of chance as probability conditional on an appropriate partition, conditionals with chance consequents,

If p, then Chance$(q) = a$.

are covered by the preceding theory. The consequent, Chance$(q) = a$, is given a truth value at each point by the construction of the random variable, Chance(q), as probability conditional on a partition.

In the general case, we may want to retain something that we have automatically with natural factors – that the notion of chance used in the consequent goes nicely with the notion of conditional chance used for the conditional. We want Chance$(q) = a$ to be equivalent to the conditional: If tautology, then Chance$(q) = a$. So we shall choose the partition used to objectify chance as $\pi(T)$, the partition that the family assigns to the tautology.

It is much easier for conditionals with chance consequents than for conditionals with unqualified consequents to have truth values. They will *always* have truth values if the family satisfies the following condition:

> *MESH:* For every consistent p and every element b of the
> partition $\pi(p)$, there is an element c of the chance
> partition $\pi(T)$ such that $b \cap p \subseteq c$.

7. ADAMS CONDITIONALS AND STALNAKER CONDITIONALS

Adams and Stalnaker approached the problem of conditionals from different directions. Adams analyzed conditionals from the perspective of probability theory; Stalnaker analyzed them from the perspective of modal logic. The two theories are so different in character that it is pleasant to uncover the connection discussed in Section 5: that *simple Stalnaker conditionals are equivalent to deterministic omonotonic Adams conditionals.*

There is more to the connection between these two approaches than this, however. Stalnaker has two theories, early and late, and there are related semantical theories of van Fraassen, Lewis, Vickers, Woodruff, and Chellas. Stalnaker (1968) bases his semantics for conditionals on a selection function that maps pairs of propositions p and possible worlds w onto possible worlds w'. For formal reasons there is an impossible world into which pairs with an inconsistent proposition get mapped. If we are concerned only with consistent propositions, as we are here, we can neglect this part of the theory. Then the Stalnaker selection function f must satisfy three constraints: (i) $f(w, p) \in p$. (ii) If $w \in p$, then $f(w, p) = w$. (iii) If $f(w, p) \in q$ and $f(w, q) \in p$, then $f(w, p) = f(w, q)$. The conditional "if p, then q" is true in world w just in case q is true in the selected world, $f(w, p)$. The first condition says that the selection function must select a world in which the antecedent is true. The second condition says that if the antecedent is already true in the base world w, then that world is selected. Stalnaker defends the third condition on the basis of the idea that the selection function should select the world *most similar* to w in which p is true. There cannot be two distinct worlds *most similar* to w.

But perhaps there is no unique most similar possible world. For this reason, van Fraassen (1976) suggested dropping condition (iii), and he called the resulting logic "weak Stalnaker logic." Let us call a selection function that meets conditions (i) and (ii) *van Fraassen*, and one that meets conditions (i), (ii), and (iii) *Stalnaker*. We know from Section 6 that any deterministic omonotonic family of partitions induces a Stalnaker selection function. Dropping the omonotonicity requirement gets us van Fraassen. *Any deterministic family of partitions induces a van Fraassen selection function.*

8. ADAMS CONDITIONALS, SET SELECTION FUNCTIONS, AND LEWIS CONDITIONALS

Vickers, Woodruff,[9] Chellas (1975), and Lewis (1974) have considered set-valued selection functions that map world–proposition pairs to sets of possible worlds. The conditional "if p, then q" is true in world w just in case q is true in all members of the selected set, $f(w, p)$. Every family of partitions induces a set selection function. To find $f(w, p)$, one finds the background partition that the family assigns to p and finds its member b that contains w. Then $f(p, w) = b \cap p$. If the conditional "if p, then q" is true in w on the set selection account, then the associated conditional chance is equal to 1 on the Bayesian account. The induced set selection function has the following "van Fraassenish" properties: (i) If $w \in p$, then $w \in f(p, w)$, and (ii) $f(p, w)$ is a subset of p. The first is a kind of "weak centering" property (Lewis, 1974). It also has the property that the selected sets form a partition: (iii) For any w and w', $f(p, w)$ and $f(p, w')$ are either equal or disjoint.

Stalnaker (pers. commun., 1984) gives a key connection between families of partitions and set selection functions induced by taking $f(p, w) = b \cap p$, as done earlier. *Every omonotonic family of partitions satisfying the MESH condition induces a Lewis set selection function with weak centering that meets condition (iii). Conversely, any Lewis set selection function with weak centering that meets condition (iii) induces an omonotonic family of partitions that satisfies MESH.* The connection between the probabilistic approach via objectifying families of partitions and the selection function approach generalizes in this way from the case of determinism to that of indeterminism.[10]

9. ITERATIONS

Iterations and embeddings of probability conditionals should be guided by probabilistic considerations. Embedding probability conditionals in truth functions produces nonsense, because probability conditionals cannot in general be said to have truth values. It is likewise nonsense for a probability conditional to be the antecedent of another probability conditional, because it does not pick out a set on which to condition. Iteration in the consequent can be given a natural probabilistic meaning for both updating and decision-making probability conditionals. The results are somewhat different in the two cases. This can be illustrated by examining the status of the export–import principle.

For the updating conditional $p \rightarrow q$, we associate the subjective conditional probability, $pr(q|p)$, for the reason that it is the probability that q would have after conditioning on p. So a natural way of extending the conditional would be to postulate that (i) the Value of a *proposition* is its probability, and (ii) the Value of a conditional $p \rightarrow q$ is the value of q after conditioning on p, providing that $p \cap q$ has positive prior probability. This has the consequence that $p \rightarrow (q \rightarrow r)$ and $(p \& q) \rightarrow r$ must have the same value.

22

The objective value of the decision-making conditional $p \Rightarrow q$ was defined as a random variable that takes at any point w in the probability space the value $\text{pr}(q | p \cap b)$, where b is the element of the background partition that contains w. This definition can be rephrased in a way that generalizes (Skyrms, 1988). The characteristic function of q, TVq, is a random variable that takes the value 1 at points where q is true and the value 0 elsewhere. Given this random variable, we define another that takes at any point w the value $E(TVq | p \cap b)$, where E is expectation and b is the element of $\pi(p)$ that contains w.

This construction can be iterated. Because the objective value of the decision-making conditional is a random variable, it can be put in place of TVq in the construction, giving a natural reading for iteration in the consequent. Thus, the objective value of the extended decision-making conditional $o \Rightarrow (p \Rightarrow q)$ at w is $E[\text{Value}(p \Rightarrow q) | o \cap c]$, where c is the member of $\pi(c)$ containing w.

When the decision-making conditional is thus extended, iteration corresponds to iterating the induced set selection function. Thus, in evaluating the conditional $o \Rightarrow (p \Rightarrow q)$ at w, we are first led to consider an expectation conditional on the set $o \cap c$, where c is the member of $\pi(c)$ containing w. At each point w' in $o \cap c$, we need to compute the value of the random variable, $\text{Value}(p \Rightarrow q)$. This leads us to consider, for each w' in $o \cap c$, the expectation of TVq conditional on the associated set $p \cap b'$, where b' is the member of $\pi(p)$ containing w'.

Thus the standard counterexamples to the export–import principle in Stalnaker-Lewis systems will also be counterexamples for iterated decision-making conditionals.

The decision-making conditional is not a proposition, and its objective value (OV) is a random variable rather than a truth value. However, specifying a value or range of values for the random variable [e.g., OV $(p \Rightarrow q) > .95$] does give us a proposition that is true at just those points where the random variable takes values within the claimed range. The act of asserting a decision-making conditional may implicitly be the act of asserting that it has high objective value. On such a reinterpretation, decision-making conditionals can sensibly be mixed with truth functions and iterated in the antecedent. There are new logical questions to investigate here, and again the pathbreaking work is by Adams (1986a).

10. Interaction

Given the foregoing, there are interactions between the updating and decision-making conditionals. Here is one example of some philosophical interest. Letting OV be objective value of the decision-making conditional, one can consider

$$[p \& \mathrm{OV}(p \Rightarrow q) > .95] \to q$$

whose value is the probability of q conditional on that set of points at which p is true and the value of the random variable that is the objective value of $p \Rightarrow q$ is greater than .95. On the model I have given, this updating conditional must have a value greater than .95, for the model assumes a finite background partition, $\pi(p)$, whose members b_i, objectify the conditional chances. Let $b_1, \ldots, b_n$ be the members in which $\mathrm{OV}(p \Rightarrow q) > .95$ is true. By definition, $\mathrm{pr}(q|p \cap b_i) > .95$ for $i = 1, \ldots, n$. By (finitely additive) elementary probability theory, $\mathrm{pr}[q|p \cap (b_1 \cup \cdots \cup b_n)]$ is a weighted average of factors of the form $\mathrm{pr}(q|p \cap b_i)$, with i ranging from 1 to n, and because each of the factors is greater than .95 by hypothesis, so is the average. Generalizing, we have

$$\mathrm{Value}\{[p \& \mathrm{OV}(p \Rightarrow q) > a] \to q\} > a$$

which is a kind of conditional Miller's principle.[11] Its philosophical interest lies in the connection that it establishes between updating and conditional chance.[12]

11. CONCLUDING REMARKS

Adams conditionals pose a rich set of philosophical questions. These include questions of their logic, their applications, and their interactions with each other, with modal concepts and Stalnaker-Lewis conditionals, and with probability theory. Some of the territory has been mapped out – mostly by Adams – but much remains to be explored. We can thank Ernest Adams for almost single-handed creation of a rich field of philosophical inquiry.

NOTES

1. With positive prior probability.
2. Those who first thought quantitatively about these matters in fact invented the probability conditional in the mathematical notation $[q|p]$.
3. Perhaps beyond the bounds where Adams would be willing to extrapolate them.
4. The justification is to be found in the representation theorems for subjective expected utility.
5. This is essentially the same value that I proposed earlier (Skyrms, 1980a, b) as the basic assertability value of the subjunctive conditional, on linguistic grounds. In natural language, the indicative conditional often corresponds to a belief-revision conditional, and the subjunctive to a decision-making condi-

tional. But here I agree with Gibbard (1981) that the difference between the updating and decision-making functions is what is of fundamental importance.

6. Or, in more complex contexts, degrees of belief conditional on a sigma algebra.

7. I have discussed the choice of the appropriate partition elsewhere (Skyrms, 1984, 1991).

8. The simplest plausible hypotheses about their connection fail. See Lewis (1976) and the chapters by Hájek and Hall in this volume.

9. Vickers and Woodruff never published, but are cited by Lewis (1974, p. 58). Lewis discusses (pp. 58–60) the relation between his system and the accounts of set selection functions.

10. Stalnaker (1980) revised his theory to take account of the possibility of ties in similarity rankings. This theory leads in a different direction. The idea is to consider a family of Stalnaker selection functions, use the Stalnaker semantics for each one, and then aggregate them by using van Fraassen's method (1966) of supervaluations. That is to say that propositions that are true under all the selection functions in the family are true in the aggregate evaluation, and likewise for falsity. A proposition whose truth value depends on the choice of selection function gets a truth-value gap in the aggregate evaluation.

There is a connection between the method of supervaluations and probabilistic methods. If one puts a probability measure over the Stalnaker selection functions in the family, then the inherited probability of the old Stalnaker conditional is 1 if the new supervaluational theory makes it true, and 0 if the new supervaluational theory makes it false. Putting a probability on the Stalnaker selection functions is not, however, equivalent to the Bayesian conditional: Suppose that p is true in base world w. Then each of the family of Stalnaker selection functions selects w as $f(p,w)$, and the Stalnaker conditional "if p, then q" will then have probability 1 or 0, depending on whether q is true in w. This enforces a kind of "strong centering" that is not part of the meaning of the decision-making conditional.

11. See Skyrms (1980a, 1988), Gaifman (1988), and van Fraassen (1989).

12. And it is interesting for the ways in which it can fail in models in which chance is objectified by a countably infinite partition but probability is not required to be countably additive.

REFERENCES

Adams, E. W. (1965) "On the Logic of Conditionals." *Inquiry* 8:166–97.

Adams, E. W. (1966) "Probability and the Logic of Conditionals." Pp. 265–316 in *Aspects of Inductive Logic*, ed. J. Hintikka & P. Suppes. North Holland, Amsterdam.

Adams, E. W. (1970) "Subjective and Indicative Conditionals." *Foundations of Language* 6:89–94.

Adams, E. W. (1975) *The Logic of Conditionals*. D. Reidel, Dordrecht.

Adams, E. W. (1976) "Prior Probabilities and Counterfactual Conditionals." Pp. 1–21 in *Foundations of Probability Theory, Statistical Inference, and Statistical Theories of Science*, vol. 1, ed. W. L. Harper & C. A. Hooker. D. Reidel, Dordrecht.

Adams, E. W. (1981) "Truth, Proof and Conditionals." *Pacific Philosophical Quarterly* 62:323–39.

Adams, E. W. (1984) "Remarks on Convention T's Pragmatic and Semantic Associations, and Its Limitations." *Pacific Philosophical Quarterly* 65:124–39.

Adams, E. W. (1986a) "Remarks on the Semantics and Pragmatics of Conditionals." Pp. 168–77 in *On Conditionals*, ed. E. Traugott et al. Cambridge University Press.

Adams, E. W. (1986b) "On the Logic of High Probability." *Journal of Philosophical Logic* 15: 255–79.

Adams, E. W. (1987) "On the Meaning of the Conditional." *Philosophical Topics* 15: 5–22.

Adams, E. W. (1988) "Consistency and Decision: Variations on Ramseyan Themes." Pp. 49–69 in *Causation in Decision, Belief Change and Statistics*. Kluwer, Dordrecht.

Chellas, B. (1975) "Basic Conditional Logic." *Journal of Philosophical Logic* 4: 133–228.

Gaifman, H. (1988) "A Theory of Higher Order Probability." Pp. 191–219 in *Causation, Chance and Credence*, ed. B. Skyrms & W. L. Harper. Kluwer, Dordrecht.

Gibbard, A. (1981) "Two Recent Theories of Conditionals." Pp. 211–47 in *Ifs*, ed. W. L. Harper, R. Stalnaker, & G. Pearce. D. Reidel, Dordrecht.

Gibbard, A. & Harper, W. L. (1978) "Counterfactuals and Two Kinds of Expected Utility." Pp. 125–62 in *Foundations and Applications of Decision Theory*, ed. C. A. Hooker, J. J. Leach, & E. F. McClennan. D. Reidel, Dordrecht.

Jeffrey, R. (1965) *The Logic of Decision*. McGraw-Hill, New York. Second edition (1983), University of Chicago Press.

Lewis, D. (1974) *Counterfactuals*. Harvard University Press, Cambridge, Mass.

Lewis, D. (1976) "Probabilities of Conditionals and Conditional Probabilities." *Philosophical Review* 85: 297–315.

Lewis, D. (1980) "Causal Decision Theory." *Australasian Journal of Philosophy* 59: 5–30.

Savage, L. J. (1954) *The Foundations of Statistics*. Wiley, New York.

Skyrms, B. (1980a) *Causal Necessity*. Yale University Press, New Haven, Conn.

Skyrms, B. (1980b) "The Prior Propensity Account of Subjunctive Conditionals." Pp. 259–65 in *Ifs*, ed. W. L. Harper, R. Stalnaker, & G. Pearce. D. Reidel, Dordrecht.

Skyrms, B. (1984) *Pragmatics and Empiricism*. Yale University Press, New Haven, Conn.

Skyrms, B. (1988) "Conditional Chance." Pp. 161–78 in *Probability and Causality*, ed. J. Fetzer. D. Reidel, Dordrecht.

Skyrms, B. (1991) "Stability and Chance." Pp. 149–61 in *Existence and Explanation*, ed. W. Spohn et al. Kluwer, Dordrecht.

Stalnaker, R. (1968) "A Theory of Conditionals." Pp. 98–112 in *Studies in Logical Theory*, ed. N. Rescher. Blackwell, Oxford.

Stalnaker, R. (1970) "Probability and Conditionals." *Philosophy of Science* 37: 68–80.

Stalnaker, R. (1980) "A Defense of Conditional Excluded Middle." Pp. 87–104 in *Ifs*, ed. W. L. Harper, R. Stalnaker, & G. Pearce. D. Reidel, Dordrecht.

Stalnaker, R. (1981) "Letter of David Lewis." Pp. 151–2 in *Ifs*, ed. W. L. Harper, R. Stalnaker, & G. Pearce. D. Reidel, Dordrecht.

van Fraassen, B. (1966) "Singular Terms, Truth-value Gaps and Free Logic." *Journal of Philosophy* 63: 481–95.

van Fraassen, B. (1976) "Probabilities of Conditionals." Pp. 261–308 in *Foundations of Probability Theory, Statistical Inference, and Statistical Theories of Science*, vol. 1, ed. W. L. Harper & C. A. Hooker. D. Reidel, Dordrecht.

van Fraassen, B. (1989) *Laws and Symmetry*. Clarendon Press, Oxford.

3

Letter to Brian Skyrms

ROBERT STALNAKER

April 3, 1984

Dear Brian:

I'm sorry I didn't get started earlier working on your very ingenious account of conditional chance. If I had I might have had something more substantive to say in my comments, and I might have been able to get you something in advance. But I have been working on it further since, and maybe I will take the opportunity at the Western Ontario meeting to talk about it (assuming that this is what you are doing there). Anyway, here are some results which help me, at least, to see more clearly what your account comes to.

First, the general family of partitions account of conditional chance is equivalent to the following set selection function account:

Let f be a function taking possible worlds and consistent propositions into consistent propositions meeting the following conditions:

(1) $$f_i(A) \subseteq A$$

(2) $$\text{If } i \in A, i \in f_i(A)$$

(3) $$f_i(A) \text{ and } f_j(A) \text{ are either disjoint or identical}$$

Conditional chance is defined in terms of degree of belief as follows:

(4) $$\text{ch}_i(A/B) = \text{pr}(A/B \cap f_i(B))$$

This formulation is equivalent to yours in that the s-function and the corresponding family of partitions are interdefinable. Any s-function model will correspond to a family of partitions model, and vice versa. The definitions of s-functions in terms of partitions and partitions in terms of s-functions are as follows:

(5) $f_i(A) = b_i^A \cap A$ (where b_i^A is the member of the A
                                      partition which contains i)

(6) $b_i^A = \{j : f_j(A) = f_i(A)\}$

You propose two constraints on the family of partitions, omonotonicity and MESH. Each corresponds to a constraint on the set selection functions. Omonotonicity corresponds to the following condition:

(7) $f_i(A \vee B) \subseteq f_i(A) \cup f_i(B)$

Omonotonicity plus MESH corresponds to the following:

(8)   If A entails B and $f_i(B)$ overlaps A, then $f_i(A) = f_i(B) \cap A$

Condition (8) is equivalent to the assumption that the selection function induces a weak total ordering of the possible worlds. So the selection function is a Lewis selection function (with weak rather than strong centering). Any Lewis selection function meeting condition (3) will induce a family of partitions satisfying omonotonicity and MESH, and any family of partitions meeting these conditions will induce a Lewis selection function.

MESH seems to me an important condition which is not given enough emphasis in your discussion. It is, I think, a kind of conservative principle: it says that you may fatten the partition cells only when it is necessary in order to make all the cells compatible with the condition. Without MESH, the conditional chance functions won't all be probability functions. With it, they always will be. (They will be Popper functions. Or are we calling them Harper functions now?)

Because of condition (3), we can say more about the similarity ordering of possible worlds that determines, or is determined by, the selection function. This ordering will always have the following features: First, worlds in the same cell of the initial partition are always equivalent with respect to similarity to any world. Second, worlds that are not in the same cell of the initial partition are never equivalent with respect to similarity to any world. In other words, the set selection function will determine a strict total ordering, relative to each cell of the partition, of all of the cells of the partition. In effect, the selection function selects, relative to i and A, the intersection of A with the closest cell to i which overlaps A.

One upshot is that your general theory is more highly constrained, and in a sense more deterministic, than at first appears. Even in the general theory there is a kind of uniqueness assumption. Although I was not as clear about this at the time, this is what the example I gave was pointing to. To repeat the example: the coin is either biased 3/1 for heads, 3/1 for tails, or fair (with equal degrees of belief in each). Now consider the conditional chance, in a possible world in which the coin is in fact fair,

28

that the coin lands heads on the condition that it is not fair. What is the right partition for that condition? It can't be the one-membered partition, since that conflicts with MESH. So it must be either the partition that groups fair with biased heads or fair with biased tails. Both seem arbitrary. If we ask what is the basic assertability value of the conditional "heads if not fair," the answer will be either 7/12 or 5/12, depending on which way we group the partitions. But, it seems to me, it should be 1/2.

Although the selection function theory is equivalent to yours – just a formulation in terms of different primitives – it does suggest different possibilities for modification and generalization. The problem suggested by my example would be avoided if one gave up condition (3), permitting selection functions which did not determine partitions. But I am not sure how to make the theory more flexible within the family of partitions framework. Giving up MESH would be too extreme, I think.

I hope this is intelligible. Maybe we can talk about it in London.

Yours sincerely,

Bob Stalnaker

4

Conditionals as random variables

ROBERT STALNAKER AND RICHARD JEFFREY

1. INTRODUCTION

Theses identifying probabilities of ordinary conditional statements with conditional probabilities of their consequents given their antecedents have seemed to many philosophers and logicians to be plausible and to promise explanations of the logic and semantics of conditional statements and of their role in inductive reasoning.[1] But the triviality theorems of David Lewis, Alan Hájek, and Ned Hall show that on general assumptions that are essentially independent of facts about actual usage, the most straightforward of these theses is untenable:

> **CCCP**: Conditional sentences have unconditional
> probabilities, which always equal the conditional
> probabilities of their consequents on their antecedents.

("CCCP" or "conditional construal of conditional probability" is Hájek and Hall's term.) Thus, Hájek (1989) shows that if the range of P is finite, there are more distinct conditional probabilities $P(B|A)$ than probabilities P (If A, then B) of conditionals. These theorems seem to substantiate the familiar warning that we come to grief if we imagine the bar in "$P(B|A)$" to be a sentential connective, "if." Understanding the bar as a mere typographical variant of a comma separating the arguments of a two-place function P, it is obvious that it makes no sense to iterate it. Then inscriptions like the following make no sense at all:

$$P((C|B)|A), \qquad P(C|(B|A))$$

In particular, they are not interpretable as probabilities of the sentences

> If A, then C if B.
>
> If B if A, then C.

31

In the notation we shall be using, these sentences become

$$A > (B > C), \qquad (A > B) > C$$

Ernest Adams' interpretation of conditional sentences escapes these triviality results, being motivated by the view of "|" as an elongated comma:

> **Adams**: Conditional sentences lack truth values and probabilities. By the "probability" of such a sentence we mean the conditional probability of its consequent on its antecedent.

That is how the term "probability" is to be understood when applied to conditionals in Adams' definition of validity for arguments with conditional premises or conclusions:

> An argument is valid iff the probability of its conclusion gets as close as you please to 1 when the probabilities of all premises get high enough (while remaining short of 1).

On this view, if we are to treat "if" as a sentence connective, we must use it only as the main connective. Forms like the two rejected earlier are ruled out, as are truth functional compounds of sentences of which some are conditionals. From this perspective, our aim in this chapter can be described as an embedding of Adams' treatment of the simplest conditionals in a more permissive or comprehensive framework allowing arbitrary embeddings of conditional sentences within each other and within truth functional compounds.

To explain this modification of Adams' thesis, we write the CCCP thesis as

$$(1) \qquad\qquad P(A > B) = P(B|A)$$

where $P(B|A) = P(A \wedge B)/P(A)$ in case $P(A) \neq 0$. Here the semantic values of the sentences $A, B, (A > B)$ that appear in the equation are propositions (i.e., the sets of possible worlds at which the sentences are true). Such propositions form a field of sets, a Boolean algebra of propositions to which some normalized measure **M** assigns the values that the function P assigns to the corresponding sentences:[2]

$$P(\text{sentence}) = \mathbf{M}(\text{the sentence's semantic value})$$

We shall follow Vann McGee in using capital roman letters, A, B, C, and so forth, as metalinguistic variables for *factual* sentences – sentence letters and their truth functional compounds – and the Greek letters φ, ψ, and so forth, for sentences generally, in which the conditional construction may occur. Our modification of the Adams thesis can be presented as a version of this $P(\text{sentence})$ formula in which the semantic value of a factual sentence

A is taken to be a function $v(A)$ assigning the value $v_\alpha(A) = 1$ to worlds α at which A is true, and assigning 0 elsewhere; think of $v(A)$ as f, and $v_\alpha(A)$ as $f(\alpha)$. In these terms, the P(sentence) formula can be written as[3]

(2) $$P(A) = \mathbf{M}\{\alpha : v_\alpha(A) = 1\}$$

The divergence from Adams' thesis comes when we replace the factual sentence A by a sentence φ that may involve the conditional construction. Here, the measure $\mathbf{M}$ be replaced by probabilistic expectation $\mathbf{E}$:

(3) $$P(\varphi) = \mathbf{E}(v(\varphi))$$

Functions like $v(A)$ – "indicator functions" assigning values 0 or 1 to possible worlds – are special sorts of *random variables* (also known as *measurable functions*). In general, random variables are certain functions assigning values to possible worlds. To qualify as a random variable, such a function must assume its values on "measurable" sets (i.e., sets to which $\mathbf{M}$ assigns values). The expectation $\mathbf{E}(f)$ of a random variable is a weighted average of the values f assumes at possible worlds – the weight of each value being the measure of the set of worlds at which f assumes it. Where the set of values actually assumed is finite, this average is a finite sum. Thus, if f is the semantic value of a factual sentence A [i.e., an indicator function $f = v(A)$], then $\mathbf{E}(f)$ or $\mathbf{E}(v(A))$ will be 1 times the measure of the set $v(A)$ of worlds at which A is true, plus 0 times the measure of the set of worlds at which A is false; that works out to be $\mathbf{M}(\{\alpha : v_\alpha(A) = 1\})$. Then for a factual sentence $\varphi = A$, our modification (3) of the Adams thesis comes to the same thing as (2). But we have not yet specified the semantic values that our modification assigns conditional sentences.

In general we take $v(\varphi)$ to be a random variable, chosen so as to satisfy the following modification or generalization of Adams' thesis:[4]

> **Generalized Adams**: Sentences have random variables as semantic values. By the "probabilities" of sentences we mean the (unconditional) expectations of those values. The values of conditional sentences with factual antecedents can be determined so as to satisfy the generalized Adams condition: equation (4).

(4) $$\mathbf{E}(v(A > \varphi)) = \mathbf{E}(v(\varphi) | \{\alpha : v_\alpha(A) = 1\})$$

In fact, there is just one way in which $v(A > \varphi)$ can be defined so as to satisfy (4), if the value of a conditional is to agree with that of its consequent wherever its antecedent is true, and have a single value wherever its antecedent is false:

(5) At each world α where A is false, $v_\alpha(A > \varphi) = \mathbf{E}(v(A > \varphi))$

33

Our aim in this chapter is to lay on the table one theoretical account of a conditional for which this modification of Adams' thesis holds and to relate it to other theoretical accounts that have been developed. We do not aim to defend the thesis as applied to natural language conditionals, but to bring out what it entails by exploring the properties of a conditional for which it holds by definition.[5]

The CCCP assumes that conditionals, like other kinds of statements, express propositions, and so say things that are true or false. Just as non-conditional sentences are true or false at each of the points (possible worlds) in the probability space, so conditional sentences will be true or false at each point, on this assumption. But what propositions do they express? What semantic rule determines the truth conditions for the conditional as a function of the values of the constituents and other features of the model? We sidestep this question. One of our conditionals assumes the same value – but not a truth value – at all worlds where its antecedent is false, that is, the value $P(\text{consequent} \mid \text{antecedent})$. Such conditionals might be called "matter-of-fact," in contrast to counterfactual conditionals, which typically assume various values – truth values, they are – at various worlds where their antecedents are false. Matter-of-fact conditionals seriously address only "factual" worlds, worlds where their antecedents are true, assuming there the same values their consequents do. Such conditionals treat all counterfactual worlds alike, assuming there whatever values they must in order to balance equation (4).

We begin as Adams did, by assuming that a conditional belief is not a categorical belief in a conditional proposition, but a qualified or conditional belief in the consequent. To have judgmental probability p for something of the form $A > B$ is not to have probability p that the world is such that $A > B$ is true, but to have probability p under the condition A that the world is such that B. Degree of belief in an unconditional sentence B is degree of belief in the truth of B, one's unconditional probability for B. Degree of belief in a conditional sentence $A > B$ is degree of belief in B conditionally on A (i.e., conditional probability for B given A). On Adams' thesis there is no further question about the semantic value of a conditional, for there are no semantic values involved other than those of the antecedent and consequent. But for us, questions arise about how conditional sentences, with random variables as semantic values, interact with each other and with other sentences in the context of reasoning.

If conditionals express propositions, then as soon as one has said what propositions they express, one has determined an interpretation for complex sentences that have conditionals as parts. But if conditionals are interpreted as random variables, then in the absence of further assumptions we have no account of complex statements with conditionals as parts. If we say that $A > B$ and $C > D$ are each "believed" when the conditional probabilities of their consequents on their antecedents are high, is $(A > B) \wedge (C > D)$ something that might be believed? And if so, what probability value must be high for it to be believed? Adams' program rejects such compounds, but Vann McGee, in a project designed to extend that program, defends an account of compounds with conditional parts. For McGee, conjunctions, negations, and disjunctions of conditionals make sense, but as with simple conditionals, to believe one of them is to have a complex attitude whose only propositional contents are truth functions of the unconditional factual components of the conditionals.

The project we explore follows a middle road between Adams' project and the propositional one. On the one hand, we follow Adams and McGee in assuming that the only propositions are nonconditional factual propositions. As with McGee, probabilities of conditionals and of all compounds of conditionals will be determined by the probabilities of such nonconditional propositions. But on the other hand we follow the propositionalists in assuming that a conditional expresses something that, like a proposition, takes a value at each possible world, or point in the probability space. And just as the probability of a proposition is the expectation, or weighted average, of its value, 0 or 1, at each possible world, so the probability of a conditional will be the weighted average of its values at the possible worlds that make up the probability space. The semantic values of conditionals, on our account, are a generalization of propositions. Propositions (i.e., functions from possible worlds into $\{1, 0\}$) constitute a special case of random variables (i.e., functions from possible worlds to real numbers). In the semantics we shall develop, the value of a conditional is a random variable with values in the interval $[1, 0]$. Specifically, it is the random variable whose value is the same as that of the consequent when the antecedent is true, and the same as the conditional expectation of the consequent on the antecedent when the antecedent is false.

This is the starting point. Our project is to extend the random variable semantics first to truth functional compounds involving simple conditionals, with factual antecedents and consequents, then to nested conditionals with propositional antecedents, and finally to a full conditional language with no restrictions on the nesting of conditionals. Even in the richest of the languages we consider, our conditional is "matter-of-fact" in the following sense: The semantic value of any sentence is a function of the truth values and probability values of sentences of the original nonconditional language. Our theory will make contact with, and draw on, several other accounts of conditionals and probability that explicate, extend, and defend Adams' thesis. In fact, we think one of the interests of our account is that it provides a common framework for a number of superficially very different ways of developing the thesis. Specifically, our basic theory will be equivalent in one sense to a fragment of McGee's theory, and equivalent in another sense to McGee's complete theory. And our final extension will be based on a construction developed some years ago by Bas van Fraassen.

2. THE BASIC THEORY

To make our successive systems precise, we define a sequence of four languages of increasing expressive power, $L_0 \subset L_1 \subset L_2 \subset L_3$. All of the languages contain sentence letters, the truth functional connectives[6] "$\sim$" and "$\wedge$," and the relevant instances of the following rule schemata:

$$\text{If } \varphi \in L_i, \text{ then } \sim \varphi \in L_i.$$

$$\text{If } \varphi, \psi \in L_i, \text{ then } (\varphi \wedge \psi) \in L_i.$$

L_0 is the smallest set of sentences meeting these conditions, that is, the language of the truth functional propositional calculus. The other three languages all use, in addition, the conditional connective "$>$" and the following formation rules:

$$\text{For } L_1: \text{ If } A, B \in L_0, \text{ then } (A > B) \in L_1.$$

$$\text{For } L_2: \text{ If } A \in L_0 \text{ and } \varphi \in L_2, \text{ then } (A > \varphi) \in L_2.$$

$$\text{For } L_3: \text{ If } \varphi, \psi \in L_3, \text{ then } (\varphi > \psi) \in L_3.$$

So L_1 allows only first-degree conditionals, conditionals with no conditional parts. L_2 allows, in addition, right-nested conditionals of any degree – conditionals that may have conditional consequents, but must have conditional-free antecedents. L_3 is the usual language of conditional logic, putting no restrictions on the nesting of conditionals.

Our semantics begins with a probability model for the language L_0 that is a quadruple $\langle K, \mathscr{F}, \mathbf{M}, v \rangle$. K is a nonempty set, the probability space, or set of possible worlds: $\mathscr{F}$ is a field of subsets of K, a set of propositions; $\mathbf{M}$ is a normalized measure on $\mathscr{F}$; the measure function is a standard one, but we require that a corresponding conditional measure $\mathbf{M}(\cdot|\cdot)$ be defined for all $G, H \in \mathscr{F}$ so as to satisfy the multiplicative law $\mathbf{M}(G|H)\mathbf{M}(H) = \mathbf{M}(G \cap H)$ and the condition that $\mathbf{M}(G|H) = 1$ if $H \subset G$. The valuation function v assigning indicator functions to sentence letters of L_0 is extended to all sentences of L_0 by the usual rules for the truth functional connectives:

$$v_\alpha(A \wedge B) = v_\alpha(A) \cdot v_\alpha(B), \qquad v_\alpha(\sim A) = 1 - v_\alpha(A)$$

where $v_\alpha(A)$ is the truth value of the sentence A at the possible world α; that is, $v_\alpha(A) = 1$ if $\alpha \in v(A)$, and $v_\alpha(A) = 0$ if $\alpha \notin v(A)$.

To give a random variable semantics for L_1, we extend the valuation

36

function, permitting values that may lie between 0 and 1 for conditionals and sentences with conditional parts, and we extend the probability function so that its values for sentences of L_1 are expectations of the random variables expressed by those sentences. The rule defining the random variable expressed by a simple conditional is as described earlier:

$$v_\alpha(A > B) = v_\alpha(B) \quad \text{if } v_\alpha(A) = 1, \qquad v_\alpha(A > B) = P(B|A) \quad \text{if } v_\alpha(A) = 0$$

But this rule is not enough to determine random variables for truth functional compounds with conditional parts, because for sentences φ, ψ that need not be factual, we cannot assume that the value $v(\varphi \wedge \psi)$ of a conjunction is a function of the values $v(\varphi)$, $v(\psi)$ of its parts, as we could earlier, where φ, ψ was A, B. It does seem natural to assume, however, that the random variables, functions that for each particular possible world take sentences into real numbers, behave like probability functions, and this assumption, conjoined with one modest additional assumption concerning truth functional compounds of conditionals, will suffice to extend the theory to L_1. The additional assumption is that the values of truth functional compounds of conditionals, like the values of simple conditionals, are the same as their overall probability in the case where the antecedents are false.

The following five conditions make our assumptions explicit:

1. If φ and ψ are tautologically equivalent, then $v_\alpha(\varphi) = v_\alpha(\psi)$.[7]
2. $0 \leqslant v_\alpha(\varphi)$, $v_\alpha(\text{tautology}) = 1$.
3. $v_\alpha(\varphi \vee \psi) = v_\alpha(\varphi) + v_\alpha(\psi) - v_\alpha(\varphi \wedge \psi)$.
4. If $\{K_1, K_2, \ldots\}$ is a partition of K, with $K_i \in \mathscr{F}$ for each i, and if for each i, $v_\alpha(\varphi) = r_i$ for all $\alpha \in K_i$, then $P(\varphi) = \sum_i \mathbf{M}(K_i) r_i$.
5. If φ is a Boolean compound of conditionals, all of whose antecedents take value 0 at α, then $v_\alpha(\varphi) = P(\varphi)$.

The first three conditions are general structural conditions governing random variables, making no specific mention of conditionals. The fourth condition requires that probabilities of sentences be the expectation values of the random variables that are their semantic values. The fifth condition, the only one that specifically concerns conditionals, is a natural extension to Boolean compounds of conditionals of the assumption for simple conditionals that the value should be the same as the conditional probability in the case where the antecedent is false. This condition makes no assumption about what the value is in this case – just that whatever it is, it is the same as the overall expectation of that random variable. Nevertheless, this assumption, together with the general structural constraints, is sufficient to determine a unique random variable value for each of the sentences of L_1 in terms of the original probability model defined on L_0.

37

Condition 5 is essentially the same as the independence assumption that McGee makes in his extension of Adams' program, and the formula entailed by these conditions for calculating the expectation of the random variable expressed by a conjunction of conditionals is exactly the same as McGee's formula for the probability of such a conjunction:[8]

(McGee) $P((A > B) \land (C > D))$

$$= \frac{P(\sim A \land C \land D)P(A > B) + P(\sim C \land A \land B)P(C > D) + P(A \land B \land C \land D)}{P(A \lor C)}$$

Call an extension of a probability model $\langle K, \mathscr{F}, \mathbf{M}, v \rangle$ a *random variable model* for $\mathbf{L}_1$ if $\mathbf{M}$ and v are extended to all sentences of $\mathbf{L}_1$ and satisfy the Generalized Adams thesis as well as the five conditions listed earlier.

Theorem 1: For any probability model $\langle K, \mathscr{F}, \mathbf{M}, v \rangle$ there is a unique random variable model for $\mathbf{L}_1$ that extends it.

Proof sketch. The laws of the probability calculus allow us to express the probability of any truth function of sentence letters and conditionals as a function of the probabilities of conjunctions of some or all of those letters and conditionals. Generalizations of McGee's formula, derivable from our assumptions, allow us to express the probabilities of conjunctions of conditionals in terms of probabilities of conjunctions having fewer conditional conjuncts, and so, iterating, reduce down to probabilities of factual sentences.[9]

In McGee's theory, as in the random variable semantics we have defined, probability values for all sentences of $\mathbf{L}_1$ are determined by the initial probability distribution on the sentences of $\mathbf{L}_0$, and the values are the same as in the random variable account. Thus McGee's theory, restricted to the language $\mathbf{L}_1$, is essentially equivalent to the random variable account, but the intuitive picture is slightly different. In the random variable picture, the probability of a conditional or a sentence containing conditionals is an expectation, a value that is derived, by averaging, from the values that the sentence receives at the particular possible worlds. On the Adams-McGee picture, the probability of a statement containing conditionals is a global feature of the probability space, representing a complex propositional attitude toward the factual propositions that are the constituents of the statement. But the values of the random variables can be recovered from the probability distribution by conditionalization. If φ is a sentence of $\mathbf{L}_1$, containing no sentence letters other than $A_1, \ldots, A_n$, then the value of the random variable at a possible world α will be the probability of $\mathscr{F}$ conditional on the conjunction $(\pm A_1 \land \cdots \land \pm A_n)$ in which each $\pm A_i$ is A_i or $\sim A_i$ depending on whether A_i is true or false in α.

3. THE FIRST EXTENSIONS

McGee's theory defines probabilities, and so, implicitly, random variables, not only for the sentences of L_1 but also for all the sentences of L_2, the language that allows right-nested conditionals such as $A > (B > C)$. But McGee's theory diverges from ours when we move to the richer language. We shall explain our way of extending the semantics to L_2, and then contrast it with McGee's. We shall call the conditional that results from our extension the Jeffrey conditional, and the one that results from the alternative extension the McGee conditional.

Because the semantics for L_1 defines random variables and probabilities for conjunctions of conditionals, and conjunctions of conditionals with factual statements, it defines conditional as well as absolute probabilities for conditionals. If $\varphi \in L_1$ and $A \in L_0$, then $P(\varphi | A)$ is already defined. So we can use the original rule for the probability of conditionals to define the random variable expressed by the conditional $A > \varphi$:

$$\text{If } v(A) = 1, \text{ then } v(A > \varphi) = v(\varphi);$$
$$\text{if } v(A) = 0, \text{ then } v(A > \varphi) = P(\varphi | A).$$

But this rule does not suffice to determine values for truth functional compounds of conditionals $(A > \varphi)$. For this we need to add the following condition:

6. If $v_\alpha(A) = 1$ and $(A > \varphi)$ occurs as a truth functional constituent of ψ, then $v_\alpha(\psi) = v_\alpha(\rho)$, where ρ is the result of replacing the conditional $(A > \varphi)$ by its consequent φ throughout ψ.

An extension of a probability model for L_0 is said to be a *random variable model* for L_2 if P and v are extended to all the sentences of L_2, and satisfy condition 6 in addition to the five conditions imposed on a random variable model for L_1.

Theorem 2: There is a unique extension of any probability model of L_0 to a random variable model for L_2.

(The proof is a minor modification of the proof for Theorem 1.)

In McGee's different extension to L_2, the Generalized Adams thesis does not hold for nested conditionals, but another principle that looks at first like just a notational variant of it does hold. Suppose we define, for any A, the probability function P_A in terms of P as the conditionalization of P on A: For any A and φ, $P_A(\varphi) =_{df} P(\varphi | A)$. What McGee's generalization requires is that for all $A \in L_0$ and $\varphi \in L_2$, $P(A > \varphi) = P_A(\varphi)$. This is equivalent to the Adams thesis where the language is L_1, in which case φ must be a sentence of L_0, but the two rules come apart when we extend the theory to L_2. This is why:

39

In effect, what our random variable semantics does is to project the global probability distribution P in a particular way onto the individual worlds that are the points in the probability space. A *proposition* is true or false at a given world independently of the weight that is assigned to that world by some probability function, and independently of what other worlds are included in the probability space. But the value of a random variable expressed by a conditional at a given world will (if the antecedent is false there) depend on the global probability function relative to which the random variable is defined. Consider two different probability functions, P and Q, defined on spaces that each contain a possible world α. If proposition B is false in α, then the random variable expressed by $B > C$ relative to Q may take a different value at α than the random variable expressed by that same conditional sentence relative to P. Specifically, suppose $Q = P_A$ is the probability function defined by conditionalizing on a factual proposition A, and suppose that φ is a sentence of $\mathbf{L}_1$. The value of the random variable expressed by φ relative to P_A may be different from the value of the random variable expressed by φ relative to P, even at possible worlds that are in both probability spaces, which is to say even in possible worlds in which A is true. So even though $P_A(B) = P(B|A) = P(A > B)$ by definition for all *factual* sentences B, it may be that $P_A(\varphi) \neq P(\varphi|A)$ for some $\varphi \in \mathbf{L}_1$. This means that if we define the value of the nested conditional $A > \varphi$ for the case where A is false in terms of $P_A(\varphi)$ rather than $P(\varphi|A)$, we get a different result.

One way to put the contrast intuitively is this: Conditionals, on the random variable account, are context-dependent, which means that they change their values when the context changes, and conditionalizing changes the context. With the Jeffrey nested conditional, $A > (B > C)$, the consequent, $(B > C)$, is interpreted relative to the original context, whereas with the McGee conditional, the same consequent, $(B > C)$, is interpreted relative to the context that would result from conditionalizing on the antecedent.

The Jeffrey conditional provides what is from an abstract point of view a simpler and more straightforward generalization of the theory, and it maintains the unrestricted Adams thesis that was our original motivation for the project – unlike the McGee theory, where, as McGee points out, the thesis holds, in general, only for first-degree conditionals. On the other hand, the McGee conditional may be a better representation of right-nested conditionals in natural language. So each of these conditionals has its interest, but we need not decide between them – we can have both by opting for the Jeffrey conditional, for the following reason: If the conditional is interpreted the McGee way, then all sentences of $\mathbf{L}_2$ are reducible to sentences of L_1. That is, everything expressible in the full McGee theory for L_2 can be expressed in the L_1 fragment, which is equivalent to the L_1 part of our theory.[10] This means that the McGee conditional is definable in terms of the Jeffrey conditional. The situation is not symmetrical. If the McGee rules for nested conditionals are adopted, the Jeffrey conditional is not equivalent to anything that can be expressed.[11]

4. THE FINAL EXTENSION

The last step in the development of our theory is an extension to the full conditional language, L_3. Before turning to this generalization, we shall make some remarks about the relation between the random variable strategy and the propositional strategy. As we have noted, the random variable functions behave exactly like probability functions; that is, for fixed possible world α, v_α is a probability function defined on all the sentences of L_2. We might try to model the random variable structure by defining a probability space, and a probability model assigning regions of the space to the sentences of the language, for each possible world α. These spaces must be defined so that if $v_\alpha(\varphi) = r$ in the original model, then the value of φ in the α space must be a region with measure r. To adopt this strategy would be to adopt, at least as a formal device, the propositional strategy for explicating the thesis that the probability of a conditional is the conditional probability. The idea is to refine the original probability space – to replace the factual possible worlds with sets of more-fine-grained possible worlds relative to which conditionals express propositions. But the point is not to hypothesize that conditionals with false antecedents really do express propositions, that there is a fact of the matter whether they are true or false; the point is only to model the random variable structure. If we can give a procedure by which such a unique refined probability space is generated from the original probability model, then we shall remain faithful to our original ambition to keep our conditional a matter-of-fact conditional in the sense that the semantic values of all the sentences of our conditional language are determined by the truth values and probability values of the sentences of the original non-conditional language.

It is clear that the measures on the new subspaces that replace the original possible worlds will all be copies of the original measure, because, except in trivial probability models, the measure is recoverable from any of the random variables. And those subspaces must also contain subspaces that reflect the whole space, all the way down. What we want is a probability space with a fractal-like structure: Look closely at any small part, at what at first seemed to be just a point, and you find a reflection of the structure of the whole. Zoom in further on a small part of that part, and you find the same thing.

A construction that Bas van Fraassen developed in his 1974 paper on probability and conditionals turns out to be exactly what we need to model the random variable structure in this way. It also provides us with an extension of the random variable semantics to L_3, and a connection with propositional semantic theories of conditionals. We shall sketch the general idea of this construction and then relate it to the random variable semantics.

In van Fraassen's construction of what he calls *Bernoulli models*, the original probability model for a language like L_0 is used to define a product model the points of which are infinite sequences of the points of the original space. A point α in the original space will correspond to the set of all infinite sequences that begin with α, and the probability value of that set will be determined by the weight of α in the original probability function. Similarly, the set of sequences that begin $\alpha\beta$ will correspond to point β in the α subspace that reflects the whole space, and so on down. To illustrate with a simple abstract example, suppose we begin with a finite space with just three possible worlds, α, β, and γ. The α region of the product space can be partitioned into three subregions, $\alpha\alpha$, $\alpha\beta$, and $\alpha\gamma$, with measures standing in the same ratio as for the measures of α, β, and γ in the whole space. These can in turn be subdivided into $\alpha\alpha\alpha$, $\alpha\alpha\beta$, $\alpha\alpha\gamma$, $\alpha\beta\alpha$, $\alpha\beta\beta$, $\alpha\beta\gamma$, $\alpha\gamma\alpha$, $\alpha\gamma\beta$, $\alpha\gamma\gamma$, and so on. The probability measure on the new space is uniquely determined.

More precisely, the Bernoulli model $\langle K^*, \mathscr{F}^*, \mathbf{M}^*, v^* \rangle$ is defined in terms of our original probability model $\langle K, \mathscr{F}, \mathbf{M}, v \rangle$ as follows: K^* is the set of all infinite sequences of members of K, and $\mathscr{F}^*$ is the closure under complement and countable union of the *generating sets*, which are defined as follows: A subset G of K^* is a generating set iff for some $G_1, \ldots, G_n \in \mathscr{F}$, the set G is the Cartesian product $G_1 \times \cdots \times G_n \times K^*$. $\mathbf{M}^*$ is then defined in terms of $\mathbf{M}$ for the generating sets as follows:

If $G = G_1 \times \cdots \times G_n \times K^*$, then $\mathbf{M}^*(G) = P(G_1) \cdot \ldots \cdot P(G_n)$.

Finally, v^* assigns a proposition to each sentence letter in accord with the following rule:

If φ is a sentence letter, then $v^*(\varphi) = v(\varphi) \times K^*$.

v^* is then extended to assign a proposition in $\mathscr{F}^*$ to each sentence of L_3 using the usual rules for the Boolean connectives and a semantic rule for the conditional that is familiar from the truth conditional semantics for conditional logic.

In the possible worlds semantics for the Stalnaker conditional, $\varphi > \psi$ is true in a possible world a if and only if ψ is true in the "closest" possible world to a in which φ is true. A model contains a selection function that selects, for each possible world and proposition, a possible world in which the proposition is true. The constraints on the selection function ensure that it determines a well ordering of the selected possible worlds with respect to each possible world.

In the pure abstract semantics, where the possible worlds are primitive points, the selection function is also primitive, but in a theory such as van Fraassen's where the possible worlds have some structure, we can say more about the ordering that is the basis for selection. The new, more-fine-grained possible worlds of the Bernoulli models are infinite sequences, and so they have parts that are also possible worlds. Each sequence, or fine-grained world, can be thought of as an infinite sequence of fine-grained worlds: If α is the nth term in the sequence a, then the subsequence headed by α obtained by lopping off the first $n - 1$ terms of a is the nth world in the sequence. Van Fraassen's theory uses this ordering as the basis for the selection function relative to which the proposition expressed by a conditional is defined: For any proposition S and world (sequence) a, $f(S, a)$ is the largest subsequence of a that is a member of S. The semantic rule for the conditional is then the standard one:

$$v^*(\varphi > \psi) = \{a : f(\varphi, a) \in v^*(\psi)\}^{12}$$

That is to say, $\varphi > \psi$ is true in a possible world a if and only if ψ is true in the closest φ world to a.

The Bernoulli model is now completely defined in terms of the original probability model, and it is easy to see that the CCCP thesis, that the probability of the conditional is the conditional probability, holds for all conditionals with factual antecedents. First, if $v_a^*(A) = 1$, then $v_a^*(A > \varphi) = v_a^*(\varphi)$, which implies that $v^*(A \wedge (A > \varphi)) = v^*(A \wedge \varphi)$. Because the propositions expressed are the same, the probabilities must be the same: $P(A \wedge (A > \varphi)) = P(A \wedge B)$. Second, if $v_a^*(A) = 0$, then it follows from the structure of the selection function that $v_a^*(A > \varphi) = v_{a-1}^*(A > \varphi)$, where $a - 1$ denotes the sequence obtained by lopping off a's head: $a - 1$ is the closest world to a, other than itself.[13] This implies that a sequence will be a member of the proposition $v^*(\sim A \wedge (A > \varphi))$ if and only if the first term of the sequence is a member of $v(\sim A)$ and the sequence that follows that term is a member of $v^*(A > \varphi)$. So by the product definition of P^*, $P^*(\sim A \wedge (A > \varphi)) = P^*(\sim A)P^*(A > \varphi)$. Now substituting in the formula $P^*(A > \varphi) = P^*(A \wedge (A > \varphi)) + P^*(\sim A \wedge (A > \varphi))$, and solving for $P^*(A > \varphi)$, one gets $P^*(A > \varphi) = P^*(A \wedge \varphi)/P^*(A) = P^*(\varphi|A)$.

The final step is to define the random variable functions in terms of the

Bernoulli model. For this we cut back down to coarse-grained possible worlds, members of K rather than K^*. The value of a random variable expressed by a sentence φ of L_3 at a coarse-grained possible world α will be the measure relative to the α subspace of K^* of the fine-grained proposition expressed by φ in the Bernoulli model. Here is one way to define that measure: First we define a function that for any $\alpha \in K$ will project $[\alpha]$, the set of all members of K^* beginning with α, one-to-one onto the whole space K^*:

For any $\alpha \in K$ and proposition $F \in \mathscr{F}^*$, let $t_\alpha(F) = _{\mathrm{df}} \{a - 1 : a \in F \cap [\alpha]\}$.

What t does is to project the part of the α subspace that intersects any proposition F onto a part of the whole space whose measure is the same as the relative measure of F relative to the α subspace. Now we can define the random variable function in terms of t:

An extension of a probability model $\langle K, \mathscr{F}, \mathbf{M}, v \rangle$ is a *random variable model for* L_3 if P and v are extended to all sentences of L_3, $P(\varphi) = P^*(\varphi)$, and $v_\alpha(\varphi) = \mathbf{M}^*(t_\alpha(v^*(\varphi) \cap [\alpha]))$, where $\mathbf{M}^*$ is the measure function in the Bernoulli model based on $\langle K, \mathscr{F}, \mathbf{M}, v \rangle$.

It is clear from the definition of $\mathscr{F}^*$ that $t_\alpha(F)$ will be a proposition if F is; so because all sentences of L_3 express propositions in the Bernoulli model, this defines v for all sentences of L_3. Because t_α maps the proposition expressed by a sentence letter A onto the whole space K^* if $v_\alpha(A) = 1$, and into the empty set if $v_\alpha(A) = 0$, this definition recaptures the original valuation function. So because there is a unique Bernoulli model for any probability model, it is clear that by this definition there is a unique random variable model extending any probability model to L_3. As we have seen, the Generalized Adams thesis for conditionals with factual antecedents holds, and it is straightforward to show that our five conditions used to define random variable models for L_1 and L_2 also hold. It follows that a random variable model for L_3 extends the model for L_2. So the van Fraassen Bernoulli model provides a representation of our L_2 random variable semantics, as well as an extension of that semantics to L_3.

NOTES

1. Concerning this last: Because the probability of something being black conditionally on it being a raven need not be the same as the probability of it not being a raven conditionally on it not being black, the hypotheses (1) "all ravens are black" and (2) "all non-black things are non-ravens" are not equivalent if the thesis holds – so that red herrings need not support (1) as strongly as black crows do.

44

2. Normalized: $\mathbf{M}(K) = 1$, where K is the set of all possible worlds. Measure: $\mathbf{M}$ is non-negative and σ-additive (i.e., the value $\mathbf{M}$ assigns to a countable union of disjoint sets is the sum of the values it assigns to those sets).

3. The measure of the set on which f assumes the value x is $\mathbf{M}(f^{-1}(x))$, i.e., $\mathbf{M}\{\alpha : f(\alpha) = x\}$.

4. The conditional expectation $E(f \mid v(A) = 1)$ of a random variable f given truth of a factual sentence A is connected to unconditional expectation by the formula $E(f \mid v(A) = 1) \cdot E(v(A)) = E(f \cdot v(A))$, where $E(v(A))$ is the probability that A is true, $\mathbf{M}\{\alpha : v_\alpha(A) = 1\}$, and $E(f \cdot v(A))$ is the expectation $E(g)$ of the function $g(\alpha) = f(\alpha) \cdot v_\alpha(A)$ that agrees with f at worlds where A is true, and has the value 0 elsewhere.

5. See Lance (1991) and Edgington (1991) for reasons why conditionals like ours and McGee's will not do as interpretations of ordinary language conditionals.

6. Other Boolean connectives such as " $\vee$ " and " $\supset$ " are defined in the usual way.

7. In determining tautological equivalence, conditionals are treated as indissoluble sentence letters; e.g., $(\sim A \wedge (A > B))$ is of form $(\sim A \wedge L)$ where the second conjunct is a sentence letter, $L = (A > B)$.

8. Two applications of the law $P(\sim \varphi \wedge \psi) = P(\psi) - P(\varphi \wedge \psi)$ reduce the right-hand side to a function of $P(A > B)$, $P(C > D)$, and probabilities of conjunctions of sentence letters.

9. But there is no simple formula for the probability $P((A > B) \wedge \varphi)$ of a conjunction of a conditional with an arbitrary sentence. The general formula for the probability of a conjunction of n conditionals would be extremely tedious to express, because each combination of truth and falsity for the antecedents needs a separate term.

10. The reduction makes use of the following three equivalence principles:

$$A > (\varphi \wedge \psi) \leftrightarrow (A > \varphi) \wedge (A > \psi), \qquad A > (\varphi \vee \psi) \leftrightarrow (A > \varphi) \vee (A > \psi),$$
$$\sim (A > \varphi) \vee (A > \bot) \leftrightarrow (A > \sim \varphi)$$

By iterated substitutions based on these, one can reduce any sentence of L_2 to a Boolean compound of sentences of the form $A_1 > (A_2 > \cdots > A_n)$ where $A_i \in L_0$ for all i. Then by the import–export principle, which is valid for the McGee conditional, a sentence of this form is equivalent to a first-degree conditional $(A_1 \wedge A_2 \wedge \cdots \wedge A_{n-1}) > A_n$ – a sentence of L_1.

11. One way to see this: Take a simple probability model with just three possible worlds, of equal probability, and factual sentences that express the eight propositions. Then all sentences of L_1 will receive one of the seven values 0, $\frac{1}{6}, \frac{1}{3}, \frac{1}{2}, \frac{2}{3}, \frac{5}{6}, 1$ – i.e., the probabilities $0, \frac{1}{3}, \frac{2}{3}$, and 1 of the factual sentences and the probabilities $P(A > B) = P(B \mid A) = \frac{1}{6}, \frac{1}{2}$, and $\frac{5}{6}$ of conditional sentences – but there is an infinite number of different values that sentences of L_2 will take if the conditional is interpreted as the Jeffrey conditional.

12. The arguments of f are a proposition, a member of $\mathscr{F}^*$, and a world, a member of K^*. If φ is a sentence of L_3, then "$f(\varphi, a)$" will abbreviate "$f(v(\varphi), a)$."

13. Except where a is a sequence consisting of an infinite repetition of a single term, so that $a - 1 = a$.

REFERENCES

Adams, Ernest (1965) "The Logic of Conditionals." *Inquiry* 8: 166–97.
Adams, Ernest (1975) *The Logic of Conditionals.* D. Reidel, Dordrecht.

45

Edgington, Dorothy (1991) "The Mystery of the Missing Matter of Fact." *The Aristotelian Society* (supplementary volume) 65: 185–209.

Hájek, Alan (1989) "Probabilities of Conditionals – revisited." *Journal of Philosophical Logic* 18: 423–8.

Hájek, Alan & Hall, Ned (this volume) "The Hypothesis of the Conditional Construal of Conditional Probability."

Hall, Ned (this volume) "Back in the CCCP."

Jeffrey, Richard (1991) "Matter-of-Fact Conditionals." *The Aristotelian Society* (supplementary volume) 65: 161–83.

Lance, Mark (1991) "Probabilistic Dependence among Conditionals." *Philosophical Review* 100: 269–76.

Lewis, David (1976) "Probabilities of Conditionals and Conditional Probabilities." *Philosophical Review* 85: 297–315.

Lewis, David (1986) "Probabilities of Conditionals and Conditional Probabilities II." *Philosophical Review* 95: 581–9.

McGee, Vann (1989) "Conditional Probabilities and Compounds of Conditionals." *Philosophical Review* 98: 485–542.

Stalnaker, Robert (1968) "A Theory of Conditionals." Pp. 98–112 in *Studies in Logical Theory*, ed. N. Rescher. Blackwell, Oxford.

van Fraassen, Bas (1976) "Probabilities of Conditionals." Pp. 261–308 in *Foundations of Probability Theory, Statistical Inference, and Statistical Theories of Science*, vol. 1, ed. W. L. Harper & C. A. Hooker. D. Reidel, Dordrecht.

5

From Adams' conditionals to default expressions, causal conditionals, and counterfactuals

JUDEA PEARL

1. PRESCRIPTIVE VERSUS DESCRIPTIVE DEFAULTS

Most real-world knowledge is communicated qualitatively, yet is processed by principles defying those of classical logic. World knowledge includes, for example, which properties are typical of objects and classes, what an agent should expect given the facts observed in the world, and how the world is expected to react to actions taken by the agent. Such expectations are usually expressed in the form of defeasible conditionals, also called *defaults*, namely, conditional sentences that tolerate exceptions. A long-standing tension exists between the logical and the probabilistic approaches to dealing with such exceptions. In the logical tradition, defeasible conditionals are interpreted as conversational conventions, as opposed to descriptions of empirical reality (McCarthy, 1986; Reiter, 1987). For example, the sentence "Birds fly" is taken to express a communication agreement such as "You and I agree that whenever I want you to conclude that some bird does not fly, I will say so explicitly; otherwise you can presume it does fly." Here the purpose of the agreement is not to convey information about the world but merely to guarantee that, in subsequent conversations, conclusions drawn by the informed will match those intended by the informer. Once the agreement is established, the meaning of the sentence acquires a prescriptive character: "If you believe that x is a bird and you have no reason to believe the contrary, then you should also believe that x flies." Neither of these interpretations invokes any statistical information about the percentage of birds that fly or any

I am grateful to Ernest Adams, Hector Geffner, Moises Goldszmidt, Daniel Lehmann, Issac Levi, David Makinson, Menachen Magidor, Paul Morris, Don Nute, Brian Skyrms, and Wolfgang Spohn for providing helpful comments on several topics in this chapter. This work was supported in part by NSF grant IRI-8821444, NSF grant IRI-9200918, Air Force grant AFOSR 900136, and State of California MICRO Grants 92-122/3.

probabilistic information about how strongly the speaker believes that a randomly chosen bird actually flies. They simply convey licenses and prescriptions for building coherent sets of epistemic beliefs, given other beliefs or lack of beliefs.

Prescriptive defaults offer a powerful language for describing patterns of arguments that underlie a reasoning process. However, as these prescriptions become more complex, they tend to behave more like procedural programs than declarative sentences in logic, forcing the speaker to guard carefully against strange, unintended consequences of slight nuances of expression, and forcing the listener to carefully analyze intricate interactions among conflicting arguments. Fortunately, the interactions become significantly simpler to understand and compute when conditionals are given a *descriptive* interpretation, conveying typical properties of the external world, to which we do not need to attribute complicated modalities such as knowing, believing, and having license to believe. For example, the probabilistic statement $P[\text{Fly}(x)|\text{Bird}(x)] = \text{High}$ (read "If x is a bird, then x probably flies") offers such a clear interpretation of "Birds fly" that it is hard to refrain from viewing such conditionals as fragments of probabilistic information, albeit subjective in nature. With such declarative statements it is easier to define how the fragments of knowledge should be put together coherently, to characterize the set of conclusions that one wishes a body of knowledge to entail, and to identify the assumptions that give rise to undesirable conclusions, if any.

The reasons are several. First, probabilistic information is, by its very nature, a declarative summarization of constraints in a world external to the speaker. As such, it is empirically testable (at least in principle), it is often shared by many agents, and the conclusions drawn from it are less subject to dispute. Second, in many cases, the ultimate aim of common conversations is the transference of probabilistic knowledge, not the speaker's pattern of presumptions (which could be whimsical). In such cases, the empirical facts that caused the agent to commit to a given pattern of presumptions are more important than the presumptions themselves, because it is those empirical facts that the listening agent is about to confront in the future. However, there are several impediments to attaching probabilistic semantics to default expressions.

To those trained in traditional logics, symbolic reasoning is the standard, and nonmonotonicity is a novelty. To students of probability, on the other hand, it is symbolic reasoning that is novel, not nonmonotonicity. Dealing with new facts that cause probabilities to change abruptly from very high values to very low values is a commonplace phenomenon in almost every probability exercise and, naturally, has never attracted special attention among probabilists. The new challenge for probabilists is to find ways of abstracting the numerical character of high and low probabilities and then casting them in linguistic terms that reflect the natural process of accepting and retracting beliefs. This challenge is especially difficult if we are to insist that, at any given time, an agent's set of beliefs is deductively closed, namely, if A is believed and B is believed, $A \wedge B$ should be believed as well. It is clear that to maintain both deductive closure and belief retraction, some major compromises must be made relative to classical probabilities, and the criterion for assessing the merits of any such compromise must be based on the quality of inferences it produces, namely, on how the inferences compare with those found in natural discourse.

In Sections 2–5 we analyze some of the most promising probability-based approaches to default reasoning, those based on the infinitesimal analysis proposed by Adams (1966) and expanded by Spohn (1988a), Geffner and Pearl (1990), Pearl (1990b), Geffner (1991), Lehmann (1989), Kraus, Lehmann, and Magidor (1990), Lehmann and Magidor (1992), and Goldszmidt and Pearl (1991a, b, 1992a, b). For uniform exposition of these developments, we shall first introduce Spohn's notation for nonstandard probabilities. In Sections 6–8 we deal with causal relationships and show that Adams' construction, coupled with Markovian conditions (encoded in a graph), can account for actions, decision-making conditionals, and counterfactuals.

2. INFINITESIMAL ANALYSIS

Spohn (1988b) has introduced a system of belief revision (called OCF, for "ordinal conditional functions") that requires only integer-value addition and yet retains the notion of conditionalization, a facility that makes probability theory context-dependent, hence nonmonotonic. Although Spohn has proclaimed OCF to be "nonprobabilistic," the easiest way to understand its power and limitations is to interpret OCF as an infinitesimal (i.e., nonstandard) analysis of conditional probabilities.

Imagine an ordinary probability function P defined over a set Ω of possible worlds (or states of the world), and let the probability $P(\omega)$ assigned to each world ω be a polynomial function of some small positive parameter ε, for example, $\alpha, \beta\varepsilon, \gamma\varepsilon^2, \ldots$, and so on. Accordingly, the probabilities assigned to any proposition A, as well as all conditional probabilities $P(A|B)$, will be rational functions of ε. Now define the OCF function $\kappa(A|B)$ as

(1) $\kappa(A|B) = $ lowest n such that $\lim_{\varepsilon \to 0} P(A|B)/\varepsilon^n$ is nonzero

49

In other words, $\kappa(A|B) = n$ iff $P(A|B)$ is of the same order as ε^n or, equivalently, $\kappa(A|B)$ is of the same order of magnitude as $[P(A|B)]^{-1}$.

If we think of n for which $P(\omega) = \varepsilon^n$ as measuring the degree to which the world ω is disbelieved (or the degree of surprise were we to observe ω), then $\kappa(A|B)$ can be thought of as the degree of disbelief (or surprise) in A given that B is true. In particular, $\kappa(A) = 0$ means that A is a serious possibility and $\kappa(\neg A) > 0$ translates to "A is believed." It is easy to verify that κ satisfies the following properties:

1. $\kappa(A) = \min\{\kappa(\omega)|\omega \vDash A\}$
2. $\kappa(A) = 0$ or $\kappa(\neg A) = 0$, or both
3. $\kappa(A \lor B) = \min\{\kappa(A), \kappa(B)\}$
4. $\kappa(A \land B) = \kappa(A|B) + \kappa(B)$

These reflect the usual properties of probabilistic combinations (on a logarithmic scale), with "min" replacing addition, and addition replacing multiplication. The result is a calculus, employing integer addition, for manipulating orders of magnitude of disbeliefs. For example, if we make the following correspondence between linguistic quantifiers and ε^n,

(2)
$$
\begin{array}{lll}
P(A) = \varepsilon^0, & A \text{ is possible,} & \kappa(A) = 0 \\
P(A) = \varepsilon^1, & A \text{ is unlikely,} & \kappa(A) = 1 \\
P(A) = \varepsilon^2, & A \text{ is very unlikely,} & \kappa(A) = 2 \\
P(A) = \varepsilon^3, & A \text{ is extremely unlikely,} & \kappa(A) = 3
\end{array}
$$

then Spohn's system can be regarded as a semiqualitative logic for reasoning about likelihood and about belief revision. It takes an initial ranking $\kappa(\omega)$ and, for any observational sentence B, permits us to determine whether a sentence A is believed by testing whether $\kappa(\neg A|B) > 0$.

The weakness of Spohn's system is that it requires the specification of a ranking function $\kappa(\omega)$ before reasoning can commence. This requires knowledge that is not readily available in common discourse. For example, we might be given the information that birds fly [written $\kappa(\neg f|b) > 0$] and no information at all about properties of non-birds, thus leaving $\kappa(f \land \neg b)$ unspecified. Hence, inferential machinery is required for drawing conclusions from a set of conditionals that only partially specifies the ranking function $\kappa(\omega)$. Such machinery is provided by Adams' conditional logic (1975), to be discussed next.

Adams' logic admits fragmentary sets of conditional sentences, treats them as constraints over properties of κ, and infers only the statements that are believed in every ranking $\kappa(\omega)$ satisfying these constraints.

50

3. ADAMS' CONSERVATIVE CORE

3.1. ε-Semantics

To connect Adams' conditionals to Spohn's ranking functions, we start, like Adams, with probability functions defined over a language L of well-formed propositional formulas (wffs). Let a *truth valuation* for L be a function t that maps the sentences in L to the set $\{1, 0\}$ (1 for "true," 0 for "false") such that t respects the usual Boolean connectives. To define a probability assignment over the sentences in L, we regard each truth valuation t as a world ω and define $P(\omega)$ such that $\sum_\omega P(\omega) = 1$. This assigns a probability measure to each sentence φ of L via $P(\varphi) = \sum_{\omega \vDash \varphi} P(\omega)$, where $\omega \vDash \varphi$ denotes that ω satisfies φ.

We now consider a set Δ of *defeasible conditionals* or *defaults*

$$(3) \qquad \Delta = \{\varphi_i \to \psi_i, \qquad i = 1, 2, \ldots, n\}$$

where φ_i and ψ_i are wffs and "$\to$" is a conditional connective. Our task is to determine what plausible conclusions can be drawn from Δ and the collection of observational facts given by a formula ϕ. Adams' seminal contribution to this problem was to interpret Δ as a set of restrictions on P, in the form of *extreme* conditional probabilities infinitesimally removed from either 0 or 1. For example, the sentence $\text{Bird}(x) \to \text{Fly}(x)$ is interpreted as $P(\text{Fly}(x)|\text{Bird}(x)) \geqslant 1 - \varepsilon$, where ε is understood to stand for a small quantity that can be made arbitrarily small, short of actually being zero. The conclusions we wish to draw from ϕ and Δ are, likewise, formulas in L that, given the restrictions Δ, are forced to acquire extremely high conditional probabilities given ϕ. In particular, a propositional formula σ will qualify as a *plausible consequence* of T, written $\phi \hspace{1pt}|\!\!\sim_\Delta \sigma$, whenever the restrictions of Δ force P to satisfy $\lim_{\varepsilon \to 0} P(\sigma|\phi) = 1$.

It is convenient to characterize the set of consequences sanctioned by this semantics in terms of the set of (ϕ, σ) pairs that are entailed by a given Δ. We call this relation *ε-entailment*,[1] formally defined as follows:

Definition 1. Let $P_{\Delta,\varepsilon}$ stand for the set of probability functions licensed by Δ for any given ε, namely,

$$(4) \quad P_{\Delta,\varepsilon} = \{P : P(\psi_i|\varphi_i) \geqslant 1 - \varepsilon \quad \text{and} \quad P(\varphi_i) > 0 \quad \text{whenever } \varphi_i \to \psi_i \in \Delta\}$$

A conditional $\phi \to \sigma$ is said to be *ε-entailed* by Δ if every distribution $P \in P_{\Delta,\varepsilon}$ satisfies $P(\sigma|\phi) = 1 - O(\varepsilon)$ [i.e., for every $\delta > 0$ there exists an $\varepsilon > 0$ such that every $P \in P_{\Delta,\varepsilon}$ will satisfy $P(\sigma|\phi) \geqslant 1 - \delta$].

In essence, this definition guarantees that an ε-entailed statement is rendered highly probable whenever all the defaults in Δ are highly

51

probable. The connection between ε-entailment and plausible consequences is simply

(5) $\qquad \phi \mathrel{\vfuzz\sim}_\Delta \sigma \quad$ iff $(\phi \to \sigma)$ is ε-entailed by Δ

3.2. Axiomatic characterization

The conditional logic developed by Adams (1975) faithfully represents this semantics by qualitative inference rules, thus facilitating derivation of new plausible consequences by direct symbolic manipulations on Δ. The essence of Adams' logic is summarized in the following set of inference rules, which were restated for default theories by Geffner and Pearl (1990).

Theorem 1. Let $T = \langle \phi, \Delta \rangle$ be a default theory where ϕ is a set of ground proposition formulas and Δ is a set of defaults; σ is a plausible consequence of ϕ in the context of Δ, written $\phi \mathrel{\vfuzz\sim}_\Delta \sigma$, iff σ can be derived from ϕ using the following rules of inference:

Rule 1 (Defaults): $(p \to q) \in \Delta \Rightarrow p \mathrel{\vfuzz\sim}_\Delta q$
Rule 2 (Logic Theorems): $p \vdash q \Rightarrow p \mathrel{\vfuzz\sim}_\Delta q$
Rule 3 (Cumulativity): $p \mathrel{\vfuzz\sim}_\Delta q, \quad p \mathrel{\vfuzz\sim}_\Delta r \Rightarrow (p \wedge q) \mathrel{\vfuzz\sim}_\Delta r$
Rule 4 (Contraction): $p \mathrel{\vfuzz\sim}_\Delta q, \quad (p \wedge q) \mathrel{\vfuzz\sim}_\Delta r \Rightarrow p \mathrel{\vfuzz\sim}_\Delta r$
Rule 5 (Disjunction): $p \mathrel{\vfuzz\sim}_\Delta r, \quad q \mathrel{\vfuzz\sim}_\Delta r \Rightarrow (p \vee q) \mathrel{\vfuzz\sim}_\Delta r$

Rule 1 permits us to conclude the consequent of a default when its antecedent is all that has been learned, and this permission is regardless of the other defaults in Δ. Rule 2 states that theorems that logically follow from a set of formulas can be concluded in any theory containing those formulas. Rule 3 – called *triangularity* by Pearl (1988) and *cautious monotony* by Lehmann and Magidor (1992) – permits the attachment of any established conclusion (q) to the current set of facts (p) without the status of any other derived conclusion (r) being affected. Rule 4 says that any conclusion (r) that follows from a fact set (p) augmented by a derived conclusion (q) also follows from the original fact set alone. Finally, Rule 5 says that a conclusion that follows from two facts also follows from their disjunction.

Some meta-theorems

T-1 (Logical Closure): $p \mathrel{\vfuzz\sim}_\Delta q, \quad p \wedge q \mathrel{\vfuzz\sim}_\Delta r \Rightarrow p \mathrel{\vfuzz\sim}_\Delta r$
T-2 (Equivalent Contexts): $p \equiv q, \quad p \mathrel{\vfuzz\sim}_\Delta r \Rightarrow q \mathrel{\vfuzz\sim}_\Delta r$
T-3 (Exceptions): $p \wedge q \mathrel{\vfuzz\sim}_\Delta r, \quad p \mathrel{\vfuzz\sim}_\Delta \neg r \Rightarrow p \mathrel{\vfuzz\sim}_\Delta \neg q$
T-4 (Right Conjunction): $p \mathrel{\vfuzz\sim}_\Delta r, \quad p \mathrel{\vfuzz\sim}_\Delta q \Rightarrow p \mathrel{\vfuzz\sim}_\Delta q \wedge r$

(Transitivity): $p \supset q, \quad q \hspace{0.1em}\mid\hspace{-0.45em}\sim_\Delta r \Rightarrow p \hspace{0.1em}\mid\hspace{-0.45em}\sim_\Delta r$
(Left Conjunction): $p \hspace{0.1em}\mid\hspace{-0.45em}\sim_\Delta r, \quad q \hspace{0.1em}\mid\hspace{-0.45em}\sim_\Delta r \Rightarrow p \wedge q \hspace{0.1em}\mid\hspace{-0.45em}\sim_\Delta r$
(Contraposition): $p \hspace{0.1em}\mid\hspace{-0.45em}\sim_\Delta r \Rightarrow \neg r \hspace{0.1em}\mid\hspace{-0.45em}\sim_\Delta \neg p$
(Rational Monotony): $p \hspace{0.1em}\mid\hspace{-0.45em}\sim_\Delta r, \quad \text{NOT}(p \hspace{0.1em}\mid\hspace{-0.45em}\sim_\Delta \neg q) \Rightarrow p \wedge q \hspace{0.1em}\mid\hspace{-0.45em}\sim_\Delta r$

The *property* of rational monotony (similar to CV of conditional logic) has one of its antecedents negated; hence, its consequences cannot be derived from Δ using the five rules of inference given earlier. Nevertheless, it is a desirable feature of a consequence relation and was proposed by Makinson (1989) as a standard for nonmonotonic logics. Rational monotony would obtain if in Definition 1 we were to limit our attention to any family of distributions P_ε that were parameterized by ε and analytic in ε (Goldszmidt, Morris, & Pearl, 1993). For example, the centroid of $P_{\Delta,\varepsilon}$ would give rise to such a family P_ε. Alternatively, rational monotony obtains if we interpret each default $p \to q$ as an OCF constraint $\kappa(q|p) < \kappa(\neg q|p)$ and select one ranking function that satisfies all these constraints (see Section 4.1) (Pearl, 1990b).

Theorem 2 (Δ-monotonicity). The inference system defined in Theorem 1 is monotonic relative to the addition of defaults; that is,

(6)   if $\phi \hspace{0.1em}\mid\hspace{-0.45em}\sim_\Delta \sigma$ and $\Delta \subseteq \Delta'$, then $\phi \hspace{0.1em}\mid\hspace{-0.45em}\sim_\Delta \sigma$

This follows directly from the fact that $P_{\Delta',\varepsilon} \subseteq P_{\Delta,\varepsilon}$, because each default statement imposes a new constraint on $P_{\Delta,\varepsilon}$. Thus, the logic is nonmonotonic relative to the addition of new facts (in ϕ) and monotonic relative to the addition of new defaults (in Δ). Full nonmonotonicity will be exhibited in Section 4, where we consider stronger forms of entailment.

3.3. Consistency and ambiguity

As a basis for automated reasoning, an important feature of the system defined by Rules 1–5 is its ability to distinguish theories portraying inconsistencies (e.g., $\langle p \to q, p \to \neg q \rangle$) from those conveying exceptions (e.g., $\langle p \wedge q, p \to \neg q \rangle$) or ambiguities (e.g., $\langle p \wedge r, p \to q, r \to \neg q \rangle$). Another important feature is its computational attractiveness.

Definition 2. Δ is said to be *ε-consistent* if $P_{\Delta,\varepsilon}$ is nonempty for every $\varepsilon > 0$; else, Δ is *ε-inconsistent*. Similarly, a set of default statements $\{S_\alpha\}$ is said to be *ε-consistent with* Δ if $\Delta \cup \{S_\alpha\}$ is ε-consistent.

Definition 3. A default statement S is said to be *ambiguous*, given Δ, if both S and its denial are consistent with Δ.

Theorem 3 (Adams, 1975). If Δ is ε-consistent, then a statement $S:p \to q$ is ε-entailed by Δ iff its denial $S':p \to \neg q$ is ε-inconsistent with Δ.

In addition to Rules 1–5 of Theorem 1, the logic possesses a systematic procedure for testing ε-consistency (hence ε-entailment) that involves a moderate number of propositional satisfiability tests.

Definition 4. Given a truth valuation t, a default statement $p \to q$ is said to be *verified* under t if t assigns the value 1 to both p and q, and it is said to be *falsified* under t if p is assigned the value 1 and q is assigned the value 0. A default statement $S:p \to q$ is said to be *tolerated* by a set Δ' of such statements if there is a t that verifies S and does not falsify any statement in Δ'.

Theorem 4 (Adams, 1975). Let Δ be a finite set of default statements. Δ is ε-consistent iff in every nonempty subset Δ' of Δ there exists at least one statement that is tolerated by Δ'.

Corollary 1 (Goldszmidt & Pearl, 1991a). ε-Consistency (hence ε-entailment) can be tested in $|\Delta|^2/2$ propositional satisfiability tests.

The procedure is simply to find a default statement that is tolerated by Δ, remove it from Δ, and repeat the process on the remaining set of statements until there are no default statements left. If this process leads to an empty set, then Δ is ε-consistent; else, it is ε-inconsistent.

If the material counterpart of $p \supset q$ for each statement $p \to q$ in Δ is a Horn expression, then consistency can be tested in time quadratic with the number of literals in Δ.

When Δ can be represented as a network of default rules, the criterion of Theorem 4 translates into a simple graphic test for consistency:

Corollary 2 (Pearl, 1987). Let Δ be a *default network*, that is, a set of default statements $p \to q$ where p is an atom and q is a literal. Δ is ε-consistent iff, for every pair of conflicting arcs $p_1 \to q$ and $p_2 \to \neg q$,

1. p_1 and p_2 are distinct, and
2. there is no cycle of positive arcs that embraces both p_1 and p_2.

Theorems 3 and 4 and their corollaries are valid only when Δ consists of purely defeasible conditionals. For mixtures of defeasible and non-defeasible conditionals, ε-consistency and ε-entailment require a slightly modified procedure (Goldszmidt & Pearl, 1991a). This procedure attributes a special meaning to a strict conditional $a \Rightarrow b$ (read "if a, then definitely b") that is different from the material implication $a \supset b$. For example, conforming to common usage of conditionals, it proclaims $\{a \Rightarrow b, a \Rightarrow \neg b\}$ as ε-inconsistent and will entail $a \Rightarrow b$ from $\neg b \Rightarrow \neg a$ but not from $\neg a$.

4. The adventurous shell

The preceding adaptation of Adams' logic of conditionals yields a system of defeasible inference with rather unique features, unshared by systems based on the prescriptive reading of defaults.

1. The system provides a formal distinction among exceptions, ambiguities, and inconsistencies and offers systematic methods of maintaining coherence.
2. Multiple extensions do not arise, and preferences among arguments (e.g., toward higher specificity) are automatically inferred from the input.

However, default reasoning requires two facilities: one forcing conclusions to be retractable in the light of new refuting evidence, the other protecting conclusions from retraction in the light of new but irrelevant evidence. Rules 1–5 excel on the first requirement, but fail on the second. For instance, if we are told that birds fly $(b \rightarrow f)$ and that Tweety is a red bird $(r \wedge b)$, the system will not conclude that Tweety flies, even though there is no sentence that in any way connects color to flying. [The opposite is true in the prescriptive approaches; they excel on the second requirement, but do not retract conclusions refuted by more specific information unless all exceptions are enumerated in advance (Reiter, 1987).]

This conservative behavior emanates from our insistence that a conclusion must attain high probability in *all* probability assignments licensed by Δ – one such assignment reflects an environment in which red birds do not fly. It is clear that if we want the system to respect the communication convention that, unless stated explicitly otherwise, properties are presumed to be *irrelevant* to each other, we need to restrict the family of probability functions relative to which inferences are ratified. In other words, we should consider only functions that minimize dependencies relative to Δ, that is, they embody dependencies absolutely implied by Δ and no others.

4.1. System-Z

One way of suppressing irrelevant properties is to restrict our attention to some "well-behaved" probability functions that comply with the constraints in Δ. This can most conveniently be done within the nonstandard analysis of Spohn (1988a, sect. 2.2), where $\kappa(\omega)$ represents the degree of surprise associated with a world ω. Translating the constraints of equation (4) to the language of nonstandard analysis yields

(7) $$\kappa(\psi_i \wedge \varphi_i) < \kappa(\neg \psi_i \wedge \varphi_i) \quad \text{if } \varphi_i \rightarrow \psi_i \in \Delta$$

55

where the κ of a formula φ is given by

$$(8) \qquad \kappa(\varphi) = \min_{\omega} \{\kappa(\omega): \omega \vDash \varphi\}$$

In the language of ranking functions, Adams' ε-entailment of $\phi \rightarrow \sigma$ is obtained via the requirement that $\kappa(\neg \sigma | \phi) > 0$ hold in *all* rankings that obey the constraints imposed by equation (7). By contrast, to suppress irrelevant properties, we can first find the minimal (hence "least surprising") κ ranking permitted by the constraints in Δ and then test whether $\kappa(\neg \sigma | \phi) > 0$ holds in this minimal κ.

Remarkably, these constraints admit a unique minimal κ ranking whenever Δ is ε-consistent. Moreover, finding this minimal ranking, which was named **Z-ranking** by Pearl (1990b), requires no more computation than testing for ε-consistency according to Corollary 1. First we identify all default statements in Δ that are tolerated by Δ, assign to them a priority of $Z = 0$, and remove them from Δ. Next we assign a Z-priority of 1 to every default statement that is tolerated by the remaining set, and so on. Continuing in this way, we form an ordered partition of $\Delta = (\Delta_0, \Delta_1, \Delta_2, \ldots, \Delta_K)$, where Δ_i consists of all statements tolerated by $\Delta - \Delta_0 - \Delta_1 - \cdots - \Delta_{i-1}$. This partition uncovers a natural priority among the defaults in Δ and represents the relative "cost" associated with violating any of these defaults, with preference given to the more specific classes. In each toleration test, we need to identify all defaults $r_i: \varphi_i \rightarrow \psi_i$ in Δ for which the formula

$$(9) \qquad \varphi_i \wedge \psi_i \wedge_{\substack{j \neq i \\ r_j \in \Delta}} \varphi_j \supset \psi_j$$

is satisfiable.

Once we establish the Z-priorities on defaults, the minimal ranking on worlds is given as follows:

Theorem 5 (Pearl, 1990b). Out of all ranking functions $\kappa(\omega)$ satisfying the constraints of equation (7), the one that achieves the lowest κ for each world ω is unique and is given by

$$(10) \qquad \kappa_0(\omega) = \min \{i: \omega \vDash (\varphi_j \supset \psi_j), Z(\varphi_j \rightarrow \psi_j) \geqslant i\}$$

In other words, $\kappa_0(\omega)$ is equal to 1 plus the Z of the highest-priority default falsified in ω. Once Z is known, the rank κ_0 of any wff ϕ is given by

$$(11) \qquad \kappa_0(\phi) = \min \left\{ i: \phi \wedge_{j: Z(r_j) \geqslant i} \varphi_j \supset \psi_j \text{ is satisfiable} \right\}$$

Given $\kappa_0(\omega)$, we can now define a useful extension of Adams' ε-entailment that has been called 1-*entailment* by Pearl (1990b).

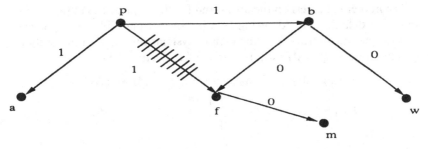

Figure 5.1

Definition 5 (1-entailment). A formula σ is said to be 1-*entailed* by ϕ, in the context Δ (written $\sigma \mathrel{\mid\!\sim}_1 \phi$), if σ holds in all minimal-κ_0 worlds satisfying ϕ. In other words,

$$(12) \qquad \phi \mathrel{\mid\!\sim}_1 \sigma \quad \text{iff} \quad \kappa_0(\phi \wedge \sigma) < \kappa_0(\phi \wedge \neg \sigma)$$

Note that ε-entailment is clearly a subset of 1-entailment.

Another important result that is implied by equations (9)–(12) gives a method of constructing a propositional theory Th(ϕ) that implies all the consequences σ that plausibly follow from a given evidence ϕ, namely, $\phi \mathrel{\mid\!\sim}_1 \sigma$. Such a theory is given by the formula

$$(13) \qquad \text{Th}(\phi) = \bigwedge_{i \,:\, Z(r_i) \geqslant \kappa(\phi)} \varphi_i \supset \psi_i$$

Clearly, if the rules in Δ are of Horn form, computing the priority Z, and therefore the ranking κ_0 of a given ϕ, can be done in polynomial time (Dowling & Gallier, 1984).

Lehmann (1988) extended ε-entailment in a different way, by syntactically closing it under the *rational monotony* rule stated earlier in the list of non-theorems, thus obtaining a new consequence relation that he called *rational closure*. Goldszmidt and Pearl (1990) showed that 1-entailment and rational closure are identical whenever Δ is ε-consistent. It is remarkable that two totally different approaches have converged on the same system of inference.

Figure 5.1 represents a knowledge base Δ containing the following defaults:

$[r_1]$ "Penguins are birds" $(p \rightarrow b)$
$[r_2]$ "Birds fly" $(b \rightarrow f)$
$[r_3]$ "Penguins do not fly" $(p \rightarrow \neg f)$
$[r_4]$ "Penguins live in the Arctic" $(p \rightarrow a)$
$[r_5]$ "Birds have wings" $(b \rightarrow w)$
$[r_6]$ "Animals that fly are mobile" $(f \rightarrow m)$

57

The numerical labels on the arcs stand for the **Z**-priorities of the corresponding defaults. The following are examples of plausible consequences that can be drawn from Δ by the various systems discussed in this section [maximum entropy (ME) will be discussed in Section 4.3]:

ε-entailed	1-entailed	ME-entailed
	$\neg b \mathrel{\mid\!\sim}_1 \neg p$	
$b \wedge p \mathrel{\mid\!\sim}_\Delta \neg f$		$p \mathrel{\mid\!\sim}^* w$
	$\neg f \mathrel{\mid\!\sim}_1 \neg b$	
$f \mathrel{\mid\!\sim}_\Delta \neg f$		$p \wedge \neg a \mathrel{\mid\!\sim}^* \neg f$
	$\neg f \mathrel{\mid\!\sim}_1 m$	
$b \mathrel{\mid\!\sim}_\Delta \neg p$		$p \wedge \neg a \mathrel{\mid\!\sim}^* w$
	$\neg m \mathrel{\mid\!\sim}_1 \neg b$	
$p \wedge a \mathrel{\mid\!\sim}_\Delta b$		
	$p \wedge \neg w \mathrel{\mid\!\sim}_1 b$	
	$r \wedge b \mathrel{\mid\!\sim}_1 f$	

1-entailment sanctions many plausible inference patterns that are not ε-entailed, such as chaining, contraposition, and discounting irrelevant features. For example, from the knowledge base of Figure 5.1 we can now conclude that birds are mobile, $b \mathrel{\mid\!\sim}_1 m$, that immobile objects are non-birds, $\neg m \mathrel{\mid\!\sim}_1 \neg b$, and that red birds fly. On the other hand, 1-entailment does not permit us to conclude that penguins who do not live in the Arctic do not fly, $p \wedge \neg a \to \neg f$.

4.2. System-Z^+

System-Z^+ is a variant of System-Z that permits defaults to be expressed with different levels of strength or conviction (Goldszmidt & Pearl, 1991b). Each default in Δ is written as a formula $\varphi \overset{\delta}{\to} \psi$, where φ and ψ are wffs and δ is a non-negative integer. The intended reading of $\varphi \overset{\delta}{\to} \psi$ is "typically, if φ, then *expect* ψ with strength δ," or $P(\psi | \varphi) \geqslant 1 - O(\varepsilon^{\delta + 1})$.[2]

The constraint conveyed by $\varphi \overset{\delta}{\to} \psi$ is the inequality $\kappa(\neg \psi | \varphi) > \delta$, which means that, given φ, it would be surprising by at least $\delta + 1$ ranks to find $\neg \psi$; it is equivalent to $\kappa(\psi \wedge \varphi) + \delta < \kappa(\neg \psi \wedge \varphi)$.

Definition 6 (Consistency). A ranking function $\kappa(\omega)$ is said to be *admissible* relative to a given Δ iff

$$(14) \qquad \kappa(\varphi_i \wedge \psi_i) + \delta_i < \kappa(\varphi_i \wedge \neg \psi_i)$$

[equivalently, $\kappa(\neg \psi_i | \varphi_i) > \delta_i$] for every rule $\varphi_i \overset{\delta_i}{\to} \psi_i \in \Delta$. A set Δ is *consistent* iff there exists an admissible ranking κ relative to Δ.

Remarkably, it can be shown that Δ is consistent iff it is ε-consistent. In other words, the δ-labels do not affect the criterion for consistency of Δ. Likewise, System-Z^+ possesses a unique minimal ranking κ^+ that depends on the δ labels.

Definition 7 (The ranking κ^+). Let $\Delta = \{r_i | r_i = \varphi_i \overset{\delta_i}{\to} \psi_i\}$ be a consistent set of defaults. κ^+ is defined as an admissible ranking function that is minimal in the following sense: Any other admissible ranking function must assign a higher rank to at least one world.

Theorem 6 (Goldszmidt & Pearl, 1992b). Any consistent Δ has a unique minimal ranking κ^+ given by

$$(15) \qquad \kappa^+(\omega) = \begin{cases} 0 & \text{if } \omega \text{ does not falsify any default in } \Delta \\ \max_{\omega \models \varphi_i \wedge \neg \psi_i}[\mathbf{Z}^+(r_i)] + 1 & \text{otherwise} \end{cases}$$

where $\mathbf{Z}^+(r_i)$ is a set of integers (*priorities*) defined on defaults that can be computed from Δ.

Thus, the default priorities $\mathbf{Z}^+$ constitute an economical way of encoding the ranking κ^+, linear in the size of Δ, from which the κ^+ of any world can be computed according to equation (15). Goldszmidt and Pearl (1991b) presented an effective procedure, Procedure Z_- rank, for computing $\mathbf{Z}^+$. Moreover, Theorem 7 and equations (11) and (13) are still valid for System-Z^+, if we substitute $\mathbf{Z}^+$ for $\mathbf{Z}$.

Theorem 7 (Goldszmidt & Pearl, 1991b). Given a consistent Δ, the computation of the $\mathbf{Z}^+$-priorities requires $O(|\Delta|^2 \times \log|\Delta|)$ satisfiability tests. Moreover, given the $\mathbf{Z}^+$-priorities, determining the ranking κ^+ of a wff ψ and the strength δ with which an arbitrary query σ is confirmed, given the information ϕ, that is, $\phi \hspace{-0.3em}\mid\hspace{-0.6em}\sim^{\delta}_+ \sigma$, requires $O(\log|\Delta|)$ satisfiability tests.

System-Z^+ can also be used to reason with *soft* evidence or imprecise observations, such as when the context ϕ of a query is not given with absolute certainty and all we have is testimony reporting that "ϕ is supported to a degree n." Goldszmidt and Pearl (1991b) studied two strategies for processing such reports. The first strategy, named *J*-conditionalization, is based on Jeffrey's rule of conditioning (Pearl, 1990a). It interprets the report as specifying that, "all things considered," the new degree of disbelief for $\neg \phi$ should be $\kappa'(\neg \phi) = n$. The second strategy, named *L*-conditionalization, is based on the *virtual evidence* proposal described by Pearl (1988). It interprets the report as specifying the desired *shift* in the degree of belief in ϕ, as warranted by that report alone and "nothing else considered." Both interpretations yield semitractable procedures (i.e., polynomial for Horn theories) for assessing the plausibility of

σ that are free from the computational difficulties that plague systems based on the prescriptive reading of defaults.

The main weakness of System-Z (and System-Z^+) is its inability to sanction property inheritance from classes to exceptional subclasses. For example, from $\Delta = \{a \rightarrow b, c \rightarrow d\}$ we cannot conclude $a \wedge \neg b \wedge c \hspace{-1pt}\mid\hspace{-4pt}\sim d$. Likewise, given the knowledge base of Figure 5.1, 1-entailment will not sanction the conclusion that penguins have wings ($p \hspace{-1pt}\mid\hspace{-4pt}\sim w$) by virtue of being birds (albeit exceptional birds). The reason is that according to System-Z all statements conditioned on p should obtain a rank of 1, and this amounts to proclaiming the penguin an exceptional type of bird in *all* respects, barred from inheriting *any* bird-like properties (e.g., laying eggs, having beaks). Sanctioning property inheritance across exceptional classes requires a more refined ordering, one that takes into account the *number* of defaults falsified in a given world, not merely their rank orders. One such refinement is provided by the maximum entropy approach (Goldszmidt et al., 1993), where each world is ranked by the sum of the weights on the defaults falsified by that world. Another refinement is provided by Geffner's conditional entailment (1991), where the priority of defaults induces a *partial* order on worlds. These two refinements will be summarized next.

4.3. The maximum entropy approach

The maximum entropy (ME) approach is motivated by the convention that, unless mentioned explicitly, properties are presumed to be independent of one another; such assumptions are normally embedded in probability distributions that, subject to a set of constraints, attain the maximum entropy (Pearl, 1988). Given a set Δ of default rules and a family of probability distributions that are admissible relative to the constraints conveyed by Δ [i.e., $P(\psi_i | \varphi_i) \geq 1 - \varepsilon, \forall r_i \in \Delta$], we single out a distinguished distribution $P^*_{\varepsilon,\Delta}$ having the greatest entropy, $-\sum_\omega P(\omega) \log(\omega)$, and define entailment relative to this distribution by

(16) $$\phi \hspace{-1pt}\mid\hspace{-4pt}\sim^* \sigma \quad \text{iff } P^*_{\varepsilon,\Delta}(\sigma | \phi) \rightarrow 1$$

Infinitesimal analysis of the ME approach also yields a ranking function $\kappa^*(\omega)$, corresponding to the lowest exponent of ε in the expansion of $P^*_{\varepsilon,\Delta}(\omega)$ into a power series in ε. It can be shown that this ranking function can be encoded parsimoniously by assigning an integer weight $Z^*(r_i)$ to each default $r_i \in \Delta$ and letting $\kappa^*(\omega)$ be the sum of the weights associated with the rules falsified by ω. The weights $Z^*(r_i)$ are governed by a set of $|\Delta|$ nonlinear equations κ_i that, under certain conditions, can be solved by iterative methods. Once the weights are established. ME-entailment is determined by the criterion of equation (16), translated to

(17) $$\phi \mathrel{\vert\!\sim}^* \sigma \quad \text{iff } \kappa^*(\neg\, \sigma \,|\, \phi) > 0$$

where

$$\kappa^*(\omega) = \sum_{i:\, \omega \models \psi_i \wedge\, \neg \phi_i} Z^*(r_i)$$

The calculation of $\kappa^*(\sigma\,|\,\phi)$ requires minimization over worlds, a task that is NP-hard even for Horn expressions (Ben-Eliyahu, 1991). In practice, however, this minimization is accomplished quite effectively in network-type databases, yielding a reasonable set of inference patterns. For example, given the knowledge base of Figure 5.1, ME-entailment will sanction the desired consequences $p \mathrel{\vert\!\sim} w$, $p \wedge \neg\, a \mathrel{\vert\!\sim} \neg\, f$, and $p \wedge \neg\, a \mathrel{\vert\!\sim} w$. Moreover, it will avoid an undesirable feature of 1-entailment that concludes $c \wedge p \mathrel{\vert\!\sim_1} \neg\, f$ from $\Delta \cup \{c \rightarrow f\}$, where c is an irrelevant property.

The ME approach has two weaknesses: It does not properly handle causal relationships, and it is sensitive to the format in which defaults are expressed. This latter sensitivity is illustrated in the following example: From $\Delta = \{$Swedes are blond, Swedes are well-mannered$\}$, ME will conclude that dark-haired Swedes are still well-mannered, whereas no such conclusion will be drawn from $\Delta = \{$Swedes are blond and well-mannered$\}$. This sensitivity might sometimes be useful for distinguishing fine nuances in natural discourse – for example, concluding that behavior and hair color are independent qualities. However, it stands at odds with most approaches to default reasoning, where $a \rightarrow b \wedge c$ is treated as shorthand for $a \rightarrow b$ and $a \rightarrow c$.

The failure to respond to causal information (Pearl, 1988, pp. 463, 519; Hunter, 1989) prevents the ME approach from properly handling tasks of temporal prediction and abduction (Hanks & McDermott, 1987), where defaults having causal character should be given priority over other defaults. This weakness is not unique to ME; it is shared by almost all descriptive and prescriptive approaches to default reasoning. A method of treating causal defaults will be presented in Section 5.

4.4. Conditional entailment

Geffner (1991) has overcome the weakness of 1-entailment by introducing two refinements. First, rather than letting rule priorities dictate a ranking function on worlds, a partial order on worlds is induced. To determine the preference between two worlds, ω and ω', we examine the highest-priority default rules that distinguish between the two worlds, namely, the rules that are falsified by one world but not by the other. If all such defaults remain unfalsified in one of the two worlds, then this world is the preferred one. Formally, if $\Delta[\omega]$ and $\Delta[\omega']$ stand for the set of defaults falsified by ω and ω', respectively, then ω is preferred to ω' (written $\omega < \omega'$) iff $\Delta[\omega'] \neq \Delta[\omega']$ and for every default r in $\Delta[\omega] - \Delta[\omega']$ there exists an r' in $\Delta[\omega'] - \Delta[\omega]$ such that r' has a higher priority than r (written $r \prec r'$). Using this criterion, a world ω will always be preferred to ω' if it falsifies a proper subset of the defaults falsified by ω'. The absence of this feature in System-Z prevents 1-entailment from concluding $p \hspace{-0.1em}\vdash\hspace{-0.5em}\sim\hspace{-0.1em} \omega$ in the example of Figure 5.1.

The second refinement introduced by Geffner is allowing the default–priority relation, "$\prec$," to become a partial order as well. This partial order is determined by the following interpretation of the default $\alpha \rightarrow \beta$: If α is all that we know, then, regardless of any other default in Δ, we are authorized to assert β. This means that $r: \alpha \rightarrow \beta$ should get a higher priority than any argument (chain of rules) leading from α to $\neg \beta$ and, more generally, that if a set of defaults $\Delta' \subset \Delta$ does not tolerate r, then at least one default in Δ' ought to have a lower priority than r. In Figure 5.1, for example, the default $r_3: p \wedge \neg f$ is not tolerated by the set $\{r_1: p \rightarrow b, r_2: b \rightarrow f\}$; hence we must have $r_1 \prec r_3$ or $r_2 \prec r_3$. Similarly, the default $r_1: p \rightarrow b$ is not tolerated by $\{r_2, r_3\}$; hence we also have $r_2 \prec r_1$ or $r_3 \prec r_1$. From the asymmetry and transitivity of "$\prec$," these two conditions yield $r_2 \prec r_3$ and $r_2 \prec r_1$. It is clear, then, that this priority on defaults will induce the preference $\omega < \omega'$, whenever ω validates $p \wedge b \wedge \neg f$ and ω' validates $p \wedge b \wedge f$; the former falsifies r_2, while the latter falsifies the higher-priority default r_3. In general, we say that a proposition g is *conditionally entailed* by f (in the context of Δ) if g holds in all the preferred worlds of f induced by every priority ordering admissible with Δ.

Conditional entailment rectifies many of the shortcomings of 1-entailment, as well as some of the weaknesses of ME-entailment. However, because conditional entailment is based on both model minimization and enumeration of subsets of defaults, its computational complexity might be overbearing. A proof theory for conditional entailment is provided by Geffner (1991).

5. WHAT'S IN AN EPISTEMIC STATE?

Adams' proclamation of conditionals as nonpropositional carriers of conditional probability information has deep repercussions on the nature of an *epistemic state*. In general, by "epistemic state" we mean a code sufficient for the determination of an agent's beliefs and changes of beliefs. A convenient assumption underlying most of probabilistic epistemology is that the state of beliefs of a rational agent can be represented by a coherent probability function P, defined over the sentences in some appropriate language. If we further invoke an infinitesimal (or order-of-magnitude) approximation, then P can be replaced by Spohn's ranking function. Alternatively, if we do not wish to commit to any possible-world construction, we can describe an epistemic state in the abstract, as a set of deductively closed beliefs and a set of transformations governing changes of those beliefs. The theory of belief revision proposed by Alchourrón, Gärdenfors, and Makinson (1985) follows this route.

In this section we shall argue that all three conceptions of epistemic states are overly simplistic because they do not account for two important modes of belief change: changes due to acquiring new conditionals, and changes due to actions and external interventions. To account for the former, we show that epistemic states must define not only beliefs and ranking of beliefs but also the identity of a finite basis of Adams' conditionals (e.g., Δ) from which beliefs and rankings emanate. To account for the latter, we show that augmenting an epistemic state with a directed acyclic graph (on atomic variables) is sufficient for constructing an adequate semantics for causal expressions, actions, and counterfactual conditionals.

5.1. Belief revision

Alchourrón, Gärdenfors, and Makinson (AGM) have advanced a set of postulates that have become the standard against which proposals for belief revision are tested (Alchourrón et al., 1985). The AGM postulates model epistemic states as deductively closed sets of (believed) sentences and impose constraints on how a rational agent should change its epistemic states when new beliefs are added, subtracted, or revised. The central result is that the postulates are equivalent to the existence of a complete preordering of all propositions according to their degree of *epistemic entrenchment*, such that belief revisions always retain more entrenched propositions in preference to less entrenched ones. Although the AGM postulates do not provide a calculus by which the revision process can be realized or the content of an epistemic state specified, they nevertheless imply that a rational revision must behave as though propositions were ordered on some scale of entrenchment.

Spohn (1988a) has shown how belief revision conforming to the AGM postulates can be embodied in the context of ranking functions. Once we specify a single ranking function $\kappa(\omega)$ on possible worlds, we associate the agent's set of beliefs with those propositions β for which $\kappa(\neg \beta) > 0$. It follows then that the models for the theory ψ representing the agent's beliefs consist of those worlds ω for which $\kappa(\omega) = 0$. To incorporate a new belief ϕ, one can raise the κ of all models of $\neg \phi$ relative to those of ϕ until $k(\neg \phi)$ becomes $\alpha \geqslant 1$, at which point the newly shifted ranking defines a new set

63

of beliefs. This process of belief revision, which Spohn named α-conditioning, represents the ranking equivalent of Jeffrey's rule of probability kinematics (Jeffrey, 1965) and was shown to comply with the AGM postulates (Gärdenfors, 1988).

In Section 4 we saw that this process of belief revision can be performed by purely syntactic operations on the conditionals in Δ, with Bayes conditioning representing $\alpha = \infty$ and J-conditionalization ($J = \alpha$) representing Jeffrey's rule. This suggests that the conditionals residing in Δ provide a sufficient representation of an epistemic state; no additional information such as world ranking or entrenchment ordering is necessary. Furthermore, as Boutilier (1992a) has shown, the epistemic entrenchment ordering of the AGM theory corresponds naturally to the Z-priorities of System-Z (more precisely, the entrenchment ordering consists in the Z-priorities of the negations of the material counterparts of the defaults in Δ).

There are, of course, computational advantages to basing the revision process on a finite set of conditional sentences, rather than on the beliefs or on the ranking or expectations that emanate from those conditionals. First, the number of propositions in one's belief set is astronomical, as is the number of worlds, whereas the number of defaults is usually manageable. Second, priorities and rankings can be extracted automatically from the content of Δ; no outside specification of belief orderings is required.

More significant, however, are the epistemological advantages of encoding belief states as sets of conditionals. This encoding permits us to respond not merely to empirical observations but also to linguistically transmitted information in the form of conditional sentences. For example, suppose someone tells us that, in addition to the information provided in Figure 5.1, "birds with no wings cannot fly." We simply add this new conditional $(b \wedge \neg w \rightarrow \neg f)$ to our knowledge base (verifying first that the addition is admissible) and recompute Z, and we then are prepared to respond to new observations or hearsay. In Spohn's system, where revisions are limited to α-conditioning, one cannot properly revise beliefs in response to conditional statements. The reason is that, given a ranking function $\kappa(\omega)$ and a conditional r, one can always find two distinct sets of conditionals, Δ_1 and Δ_2, both inducing $\kappa(\omega)$, such that $\Delta_1 \cup \{r\}$ induces a different ranking than $\Delta_2 \cup \{r\}$. For example, both $\Delta_1 = \{a \rightarrow b\}$ and $\Delta_2 = \{\neg b \rightarrow \neg a\}$ induce the same (minimal) ranking function $\kappa_0(\omega)$, yet $\Delta_1 \cup \{a \rightarrow \neg b\}$ is inconsistent, whereas $\Delta_2 \cup \{a \rightarrow \neg b\}$ is consistent. The AGM theory, too, is inadequate for describing revision due to incorporation of new conditionals. For example, Gärdenfors' attempt to devise revision postulates for conditional sentences compatible with the Ramsey test led to triviality results (Gärdenfors, 1988, pp. 156–60).

In summary, the operation of acquiring a new conditional sentence

64

cannot be encoded in terms of a transformation on ranking functions or belief sets; it requires knowing the set of conditionals that gave rise to those rankings or beliefs.

5.2. Iterated conditionals

The ability to acquire new conditionals (or defaults) also provides a simple semantics for interpreting iterated conditionals (e.g., "If you wear a helmet whenever you ride a motorcycle, then you won't get hurt badly if you fall"[3]). Iterated conditionals cease to be a mystery (Levi, 1988) once we permit explicit reference to default expressions. The sentence "If $(a \to b)$, then $(c \to d)$" is interpreted as

"If I add the default $a \to b$ to Δ, then the conditional $c \to d$ will be satisfied by the consequence relation '$\vdash$' of the resulting knowledge base $\Delta' = \Delta \cup \{a \to b\}$."

This is clearly a proposition that can be tested in the language of default-based ranking systems. Note the semantic distinction between having a conditional sentence $a \to b$ encoded explicitly in Δ and having that conditional sentence satisfied in the consequence relation of Δ. In the former case $a \to b$ is ε-entailed from Δ, whereas in the latter case it is inferred from Δ by weaker notions of entailment and hence is more vulnerable to contradicting information. This distinction gets lost in systems that do not acknowledge defaults as the basis for rankings and beliefs, and no algebraic theory of iterated conditionals can replenish this information.

There is a natural explanation for why conditionals, in addition to being assertable in some state of propositional beliefs, should also be encoded explicitly. Conditionals carry stable domain knowledge, whereas propositions carry transitory factual information. Domain knowledge specifies the tendency of things to happen (i.e., relations that hold true in all worlds), whereas propositional beliefs describe that which actually happened in one particular world.

In Section 3 we saw, for example, that it makes a profound difference whether the sentence "All penguins are birds" is treated as a conditional sentence $p \to b$ in the knowledge base Δ or as a propositional formula $\neg p \lor b$ in ϕ (read "It has been observed that this object is definitely either a non-penguin or a bird"). The former is treated as a constraint that shapes the set of admissible probability distributions (or κ rankings), whereas the latter serves as evidence upon which the admissible distributions are to be conditioned. The former gives the intended results, properly treating penguins as a subclass of birds. The latter does not, because the observation $\neg p \lor b$ can be totally subsumed by other observations, say

65

$p \wedge b$, thus yielding identical conclusions regardless of whether penguins are a subclass of birds or birds are a subclass of penguins.

6. CAUSAL CONDITIONALS, NETWORKS, AND ACTIONS

In natural discourse, conditionals often convey causal dependency between the antecedent and the consequent. Yet, strangely, none of the conditional logics proposed in the literature endows conditionals with properties we normally associate with causal dependencies, including the Stalnaker–Lewis logic of subjunctive and counterfactual conditionals. In fact, the Stalnaker–Lewis logic coincides with Adams' logic for indicative conditionals, an enigma that, aside from being a dormant embarrassment (Gibbard, 1981), clearly indicates that none of these logics has succeeded in capturing the distinct features of causal dependence. When researchers in artificial intelligence encountered difficulties in automating causal inferences, the problem of causal conditionals again became a center of attention (Hanks & McDermott, 1987; Pearl, 1988, pp. 509–16).

The problem of causal conditionals plagues both the prescriptive and the descriptive approaches to conditionals. For example, the prescriptive approach fails to block the chaining of the following default expressions:

1. If the grass is wet, then the sprinkler must have been on.
2. If I break this bottle, the grass will get wet.

We do not wish to conclude from these two expressions that if I break this bottle, the sprinkler must have been on. The descriptive approach also produces undesirable effects. Consider, for example, the following conditionals:

3. If I turn the ignition key, the car will start $(k \rightarrow cs)$.
4. If I turn the ignition key and the battery is dead, the car will not start $(k \wedge bd \rightarrow \neg cs)$.

From these two conditionals, Adams' logic and all the entailment relationships discussed in Sections 1–4 produce the following pair of inferences:

$$k \mathrel{\vdash\mkern-9mu\sim} \neg bd$$

$$\neg k \mathrel{\vdash\mkern-9mu\sim} bd \vee \neg bd$$

Taken together, these inferences seem to imply that the act of turning the ignition key has a mysterious influence over the state of the battery: The battery is believed to be OK when we turn the ignition key on, but may become defective each time we turn the ignition key off. Such behavior is counterintuitive. Regardless of our perception of the causal connection between the ignition key and the battery, we certainly do not expect that

66

stating conditionals 3 and 4, which merely predict the behavior of the car under various key/battery conditions, will introduce this connection. If the battery is presumed unaffected by the ignition key prior to describing the conditions necessary for starting the car, then it ought to remain unaffected by the ignition key after describing those conditions. This is a prevailing general pattern of causal reasoning: Contemplating possible developments of future events should not affect our belief in past and present events.

To incorporate this pattern of inference into the language of ranking functions, it is convenient to characterize the conditionals in Δ in terms of a graph or network $\Gamma(\Delta)$, where each node in Γ corresponds to an atomic variable and in which an arc is drawn from node X_i to node X_j just in case there is a conditional in Δ such that X_i appears in the antecedent and X_j appears in the consequent.[4] Goldszmidt and Pearl (1992a) have shown that in order to account for the distinct character of causal conditionals we must impose a Markovian condition on the admissible ranking functions. In other words, the definition of entailment must now invoke only ranking functions that both satisfy the constraint of equation (7) and are *stratified* relative to the network $\Gamma(\Delta)$, namely, the ranking is organized in layers such that the addition of each new layer does not introduce new dependences among variables in existing layers. Formally, stratification can be defined through a decomposition paralleling that of Bayesian networks (Pearl, 1988):

Definition 8. A ranking function $\kappa(\omega)$ is said to be *stratified* relative to a directed acyclic graph (dag) Γ if

$$(18) \qquad \kappa(\omega) = \sum_i \kappa(X_i(\omega)|\mathbf{pa}_i(\omega))$$

where $\mathbf{pa}_i(\omega)$ are the parents of X_i in Γ evaluated at state ω.

It can be shown that in a stratified ranking, each variable is conditionally independent of all its non-descendants, given its parents – a Markovian condition typical of causal influences (Suppes, 1970; Spohn, 1980; Pearl, 1988; Pearl & Verma, 1991).

Goldszmidt and Pearl (1992a) have shown that entailment relations characteristic of causal prediction and abduction indeed result from imposing the stratification constraint on the ranking functions in equation (12). This suggests that a graphic code in the form of a dag Γ should be admitted as part of an agent's epistemic state. But the role that networks play in the representation of actions provides a more compelling reason.

A *causal network* is a dag in which each node corresponds to an atomic variable and each edge $X_i \rightarrow X_j$ asserts that X_i has a direct causal influence on X_j. Such networks provide a convenient data structure for encoding two types of information: how the initial ranking function $\kappa(\omega)$ is formed and restored (when observations fade), and how external actions will influence the agent's belief ranking $\kappa(\omega)$. Formally, causal networks are defined in terms of stratification and actions.

Definition 9. A dag Γ is said to be a causal network of $\kappa(\omega)$ if $\kappa(\omega)$ is stratified relative to Γ and, in addition, represents the effect of actions as Bayesian conditioning on variables added to Γ as root nodes.

The effect of an atomic action $\text{do}(X_i = \text{true})$ is represented by adding to Γ a link $\text{DO}_i \to X_i$, where DO_i is a new variable taking values in $\{\text{do}(x_i), \text{do}(\neg x_i), \text{idle}\}$ and x_i stands for $X_i = \text{true}$. Thus, the new parent set of X_i in the augmented network is $\mathbf{pa}'_i = \mathbf{pa}_i \cup \{\text{DO}_i\}$, and it is related to X_i by

$$(19) \quad \kappa(X_i(\omega) \mid \mathbf{pa}'_i(\omega)) = \begin{cases} \kappa(X_i(\omega) \mid \mathbf{pa}_i(\omega)) & \text{if } \text{DO}_i = \text{idle} \\ \infty & \text{if } \text{DO}_i = \text{do}(y) \text{ and } X_i(\omega) \neq y \\ 0 & \text{if } \text{DO}_i = \text{do}(y) \text{ and } X_i(\omega) = y \end{cases}$$

The effect of performing action $\text{do}(x_i)$ is to transform $\kappa(\omega)$ into a new belief ranking, $\kappa_{x_i}(\omega)$, given by

$$(20) \quad \kappa_{x_i}(\omega) = \kappa'(\omega \mid \text{do}(x_i))$$

where κ' is the ranking dictated by the augmented network $\Gamma \cup \{\text{DO}_i \to X_i\}$ and equations (18) and (19).

This yields a simple and direct transformation of pre-action and post-action rankings:

$$(21) \quad \kappa_{x_i}(\omega) = \begin{cases} \kappa(\omega) - \kappa(x_i \mid \mathbf{pa}_i(\omega)), & \omega \vDash x_i \\ \infty, & \omega \vDash \neg x_i \end{cases}$$

which ensures the following aspects of actions:

1. An action $\text{do}(x_i)$ can affect only the descendants of X_i in Γ, and
2. Every nonempty set of variables $S = \{X_j : 1 \leqslant j \leqslant m\}$ contains at least one variable X_i that has no descendants in S; hence $\kappa_{x_i}(X_1, \ldots, X_{i-1}, X_{i+1}, \ldots, X_m) = \kappa(X_1, \ldots, X_{i-1}, X_{i+1}, \ldots, X_m)$. In other words, the action $\text{do}(x_i)$ does not affect any of the variables in S. This property can in fact be taken as a basic axiom for defining the concept of an *atomic* action $\text{do}(x_i)$, from which the structure of Γ can be uncovered.

7. COMBINING ACTIONS AND OBSERVATIONS

Equation (21) was derived under the assumption that $\kappa(\omega)$ is given by the sum of equation (18), which reflects our initial beliefs prior to making any observations. We now extend equation (21) to include beliefs resulting from observations.

Let $\kappa_{A|C}(\omega)$ stand for the ranking function that will prevail if we take action A after observing C; in other words, $\kappa_{A|C}(\omega)$ is the A-update of $k(\omega|C)$. To define the proper transformation from $\kappa(\omega)$ to $\kappa_{A|C}(\omega)$, we must consider two epistemic states, before and after the action, and invoke a persistence model to determine which beliefs will persist and which will be "clipped" by the influence of action A.

Such a model, as previously invoked (Pearl, 1993), assumes that persistence forces yield to causal forces, namely, that only those properties should persist that are not under any causal influence to terminate. In terms of ranking functions, this assumption yields the following result:

Theorem 8. Let A be a conjunction of atomic actions, $A = \wedge_{j \in J} A_j$, where each A_j stands for either $do(x_j)$ or $do(\neg x_j)$, and let $\kappa(\omega|C)$ be the belief ranking prior to taking action A. Then the post-action ranking $\kappa_{A|C}(\omega)$ is given by the formula

$$(22) \qquad \kappa_{A|C}(\omega) = \kappa(\omega) - \sum_{i \in J \cup R} \kappa(X_i(\omega)|\mathbf{pa}_i(\omega))$$

$$+ \min_{\omega'} \left[\sum_{i \notin J} S_i(\omega, \omega') + \kappa(\omega'|C) \right]$$

where R is the set of root nodes of Γ and

$$(23) \ S_i(\omega, \omega') = \begin{cases} s_i & \text{if } X_i(\omega) \neq X_i(\omega') \text{ and } X_i \in R \\ s_i & \text{if } X_i(\omega) \neq X_i(\omega'), \ X_i \notin R, \text{ and } \kappa(\neg X_i(\omega)|\mathbf{pa}_i(\omega)) = 0 \\ 0 & \text{otherwise} \end{cases}$$

$S(\omega, \omega')$ represents persistence assumptions: It is surprising (to degree $s_i \geq 1$) to find X_i change from its pre-action value of $X_i(\omega')$ to a post-action value of $X_i(\omega)$ if there is no causal reason for the change.

Equation (22) demonstrates that belief changes due to long streams of observations and actions can be computed as successive updating operations on epistemic states, these states being organized by a fixed causal network, in which the only varying element is the belief ranking κ.

8. COUNTERFACTUAL CONDITIONALS

Stalnaker (1981) was the first to make the connection between actions and counterfactual statements, and he proposed using the probability of the counterfactual conditional (as opposed to the conditional probability, which is more appropriate for indicative conditionals) in the calculation of expected utilities. Stalnaker's theory does not provide an explicit connection between subjunctive conditionals and causation, however. Although the selection function used in the Stalnaker-Lewis nearest-world semantics can be thought of as a generalization of, and a surrogate for, causal knowledge, it is *too* general, as it is not constrained by basic features of causal relationships such as asymmetry, transitivity, and complicity with temporal order. To obtain a more faithful logic of counterfactuals, we must be able to translate causal conditionals into specifications of the Stalnaker-Lewis selection function.[5] Such specifications were partially provided by Goldszmidt and Pearl (1992a), through an imaging function ω^*, and are further refined by invoking the persistence model of equation (22). Note that a dag is the only ingredient one needs to add to the traditional notion of *epistemic state* so as to specify a causality-based selection function.

From this vantage point, the transformation of equation (22) provides, in essence, a new account of subjunctive conditionals that is more reflective of those used in decision making. This account treats three-argument subjunctives of the form "If it were A, then B, given C," written $A > B|C$. Subjunctives of this form have the following causal interpretation: "Given C, if I were to perform action A, then I believe B would come about." In the language of ranking functions, this interpretation reads

(24) $A > B|C$ is assertable iff $\kappa(\neg B|C) = 0$ and $\kappa_{A|C}(\neg B) > 0$

The equality states that $\neg B$ is considered a serious possibility prior to performing A, whereas the inequality renders $\neg B$ surprising after performing A. Notice that $A > B|C$ cannot be reduced to $(C \wedge A) > B$ or to $C \wedge (A > B)$, because neither of these expressions reveals the subordinate role of C, when it is incompatible with either A or B.

This can be seen more clearly in the following dialogue:

Robot 1: It is dark in here.
Robot 2: Had the switch been up, the light would be on.
Robot 1: The switch *is* up.
Robot 2: Had the switch been down, the light would be on.

Robot 2's subjunctive sentences correspond to the expressions

$$UP > ON|OFF$$
$$(\neg UP > ON)|OFF \wedge UP$$

respectively, which do not admit reduction of the conditioning formulas.

The account given in equation (24), which we can call *decision-making conditionals* (DMC), avoids the CS and CSO paradoxes of conditional logics (Nute, 1992) by ratifying only those conditionals $A > B$ that reflect a causal rather than an accidental connection between A and B and by insisting that causal connections are antisymmetric.[6] This account also explains (see the foregoing dialogue) why the assertability of counterfactual conditionals is often dependent upon previous observations, a point noted by Adams (1976) and later explained in terms of probabilities of propensities by Skyrms (1981). Such propensities are now given a concrete embodiment in the form of the causal network Γ (see also Balke and Pearl, 1994).

Although transitivity is a characteristic feature of causation, the transitive rule

$$(25) \qquad [(A > B) \wedge (B > C)] \Rightarrow (A > C)$$

is certainly not supported unconditionally by all causal models. For example (Nute, 1992), George's health would improve (C) if he stopped smoking (B), and George would stop smoking (B) if he contracted emphysema (A), but surely George's health is not improved if he contracts emphysema. Indeed, DMC sanctions only weaker forms of transitivity, such as

$$[(A > B) \wedge (B > C)] \Rightarrow \neg (C > A)$$

or

$$(26) \qquad [(A > B) \wedge (B > C) \wedge \neg (A \wedge B > \neg C)] \Rightarrow (A > C)$$

But the main advantage of the decision theoretic account is its uncovering of the interrelationships between counterfactual conditionals and indicative conditionals. An example of such relationship, which is a theorem in DMC, is the so-called Reichenbach principle (Reichenbach, 1956):

$$(27) \qquad (A \to B) \wedge (\neg A \to \neg B) \wedge \neg (A > B) \wedge \neg (B > A)$$
$$\Rightarrow (C > A) \wedge (C > B) \text{ for some } C$$

where $A \to B$ stands for the indicative conditional defined by $\kappa(\neg B | A) > 0$. This principle states that every dependence must have a causal explanation, either direct or indirect (via some common cause C) (Pearl & Verma, 1991).

Notes

1. Adams (1975) named this p-entailment. However, ε-entailment better serves to distinguish this from weaker forms of probabilistic entailment (see Section 4).
2. The special case of $\delta = \infty$ corresponds to a *strict* conditional, to be denoted by "$\Rightarrow$."
3. An example due to Philip G. Calabrese (pers. commun.).
4. If Δ contains only causal conditionals, it is reasonable to assume that Γ is a directed acyclic graph (dag). Moreover, the variables $X_1, X_2, \ldots, X_n$ are "natural partitions" that, for simplicity, are assumed to be bi-valued, namely, each X_i takes value in $\{x_i, \neg x_i\}$. Thus, conditionals 3 and 4 would be represented by the network $\Gamma = K \to CS \leftarrow BD$ (capital letters designate variable names).
5. Gibbard and Harper (1981) developed a quantitative theory of rational decisions

that is based on Stalnaker's suggestion and explicitly attribute causal interpretation to counterfactual conditionals. However, they assume that probabilities of counterfactuals are given in advance and do not specify either how such probabilities are encoded or how they relate to probabilities of ordinary propositions.

6. CS stands for $A \wedge B \Rightarrow (A > B)$, according to which the outcomes of past U.S. presidential elections, for example, would lead us to conclude that "If Nixon had been elected president in 1972, then Betty Ford would have lived in the White House in 1974" (Nute, 1992, p. 856). CS is not valid in our account because it does not satisfy equation (24) for all epistemic states, e.g., when Γ is empty. CSO stands for

$$[(A > B) \wedge (B > A)] \Rightarrow [A > C) = (B > C)]$$

which, paradoxically, leads one to conclude from "If the bulb were lit (A), the battery would be good (C)" that "If the switch were on (B), the battery would be good (C)." Although CSO is a theorem in the DMS account, it is a vacuous theorem; the antecedents $(A > B) \wedge (B > A)$ could not both hold at the same time, because of causal antisymmetry. [Indeed, it is not quite right to say that by making the bulb light (A), we cause the switch to turn on (B). The bulb can be made to light by other means, e.g., by short-circuiting the switch.]

References

Adams, E. W. (1966) "Probability and the Logic of Conditionals." Pp. 265–316 in *Aspects of Inductive Logic*, ed. J. Hintikka & P. Suppes. North Holland, Amsterdam.

Adams, E. (1975) *The Logic of Conditionals*. D. Reidel, Dordrecht.

Adams, E. W. (1976) "Prior Probabilities and Counterfactual Conditionals." Pp. 1–21 in *Foundations of Probability Theory, Statistical Inference, and Statistical Theories of Science*, vol. 1, ed. W. L. Harper & C. A. Hooker. D. Reidel, Dordrecht.

Alchourrón, C., Gärdenfors, P., & Makinson, D. (1985) "On the Logic of Theory Change: Partial Meet Contraction and Revision Functions." *Journal of Symbolic Logic* 50: 510–30.

Balke, A. & Pearl, J. (1994) "Probabilistic Evaluation of Counterfactual Queries." To appear in *Proceedings of the Twelfth National Conference on Artificial Intelligence (AAAI-94)*, Seattle, WA, July 1994.

Ben-Eliyahu, R. (1991) "NP-Complete Problems in Optimal Horn Clauses Satisfiability." UCLA, Cognitive Systems Laboratory, Technical Report R-158.

Boutilier, C. (1992a) "Conditional Logics for Default Reasoning and Belief Revision." Ph.D. dissertation, University of Toronto.

Boutilier, C. (1992b) "What Is a Default Priority?" in *Proceedings of the Canadian Conference on Artificial Intelligence (CCAI-92)*, pp. 140–7. Vancouver.

Dowling, W., & Gallier, J. (1984) "Linear-Time Algorithms for Testing the Satisfiability of Propositional Horn Formulae." *Journal of Logic Programming* 3:267–84.

Gärdenfors, P. (1988) *Knowledge in Flux: Modeling the Dynamics of Epistemic States*. MIT Press, Cambridge, Mass.

Geffner, H. A. (1991) *Default Reasoning: Causal and Conditional Theories*. MIT Press, Cambridge, Mass.

Geffner, H., & Pearl, J. (1990) "A Framework for Reasoning with Defaults." Pp. 69–87 in *Knowledge Representation and Defeasible Reasoning*, ed. H. Kyburg, R. Loui, & G. Carlson. Kluwer, Dordrecht.

72

Gibbard, A. (1981) "Two Recent Theories of Conditionals." Pp. 211–47 in *Ifs*, ed. W. L. Harper, R. Stalnaker, & G. Pearce. D. Reidel, Dordrecht.

Gibbard, A., & Harper, L. (1981) "Counterfactuals and Two Kinds of Expected Utility." Pp. 153–90 in *Ifs*, ed. W. L. Harper, R. Stalnaker, & G. Pearce. D. Reidel, Dordrecht.

Goldszmidt, M., Morris, P., & Pearl, J. (1993) "A Maximum Entropy Approach to Non-monotonic Reasoning." *IEEE Transactions on Pattern Analysis and Machine Intelligence* 15: 220–32.

Goldszmidt, M., & Pearl, J. (1990) "On the Relation between Rational Closure and System-Z." Pp. 130–40 in *Third International Workshop on Nonmonotonic Reasoning*, May 1990. South Lake Tahoe, Calif.

Goldszmidt, M., & Pearl, J. (1991a) "On the Consistency of Defeasible Databases." *Artificial Intelligence* 52: 121–49.

Goldszmidt, M., & Pearl, J. (1991b) "System Z^+: A Formalism for Reasoning with Variable Strength Defaults." Pp. 399–404 in *Proceedings of the Ninth National Conference on Artificial Intelligence*, AAAI Press/The MIT Press, Menlo Park, Calif.

Goldszmidt, M., & Pearl, J. (1992a) "Rank-based Systems: A Simple Approach to Belief Revision, Belief Update, and Reasoning about Evidence and Actions." Pp. 661–72 in *Proceedings of the Third International Conference on Knowledge Representation and Reasoning*, ed. B. Nebel, C. Rich, & W. Swartout, Cambridge, Mass., October. Morgan Kaufmann, San Mateo, Calif.

Goldszmidt, M., & Pearl, J. (1992b) "Reasoning with Qualitative Probabilities Can Be Tractable." Pp. 112–20 in *Proceedings of the Eighth Conference on Uncertainty in Artificial Intelligence*, ed. D. Dubois, M. P. Wellman, B. D'Ambrosio, & P. Smets, Stanford, Calif. Morgan Kaufmann, San Mateo, Calif.

Hanks, S., & McDermott, D. (1987) "Nonmonotonic Logics and Temporal Projection." *Artificial Intelligence* 33: 379–412.

Harper, L., Stalnaker, R., & Pearce, G. (eds.) (1981) *Ifs*. D. Reidel, Dordrecht.

Hunter, D. (1989) "Causality and Maximum Entropy Updating." *International Journal of Approximate Reasoning* 3: 87–114.

Jeffrey, R. (1965) *The Logic of Decisions*. McGraw-Hill, New York. 1965. Second edition (1983), University of Chicago Press.

Katsuno, H., & Mendelzon, A. O. (1991) "On the Difference between Updating a Knowledge Base and Revising It." Pp. 387–94 in *Principles of Knowledge Representation and Reasoning: Proceedings of the Second International Conference on Principles of Knowledge Representation and Reasoning*, ed. J. Allen, R. Fikes, & E. Sandewall. Morgan Kaufmann, San Mateo, Calif.

Kraus, S., Lehmann, D., & Magidor, M. (1990) "Nonmonotonic Reasoning, Preferential Models and Cumulative Logics." *Artificial Intelligence* 44: 167–207.

Lehmann, D. (1989) "What Does a Conditional Knowledge Base Entail?" Pp. 212–22 in *Proceedings of the First International Conference on Principles of Knowledge Representation and Reasoning (KR '89)*, ed. R. J. Brachman, H. J. Levesque, & R. Reifer. Morgan Kaufmann, San Mateo, Calif.

Lehmann, D., & Magidor, M. (1992) "What Does a Conditional Knowledge Base Entail?" *Artificial Intelligence* 55: 1–60.

Levi, I. (1988) "Iteration of Conditionals and the Ramsey Test." *Synthese* 76: 49–81.

McCarthy, J. (1986) "Applications of Circumscription to Formalizing Common-Sense Knowledge." *Artificial Intelligence* 28: 89–116.

Makinson, D. (1989) "General Theory of Cumulative Inference." Pp. 1–18 in *Non-monotonic Reasoning*, vol. 346, ed. M. Reinfrank, J. de Kleer, M. Ginsberg, & E. Sandewall. Springer-Verlag, Berlin.

Nute, D. (1992) "Logic, Conditional." Pp. 854–60 in *Encyclopedia of Artificial Intelligence*, 2nd ed., ed. Stuart C. Shapiro. Wiley, New York.

Pearl, J. (1987) "Deciding Consistency in Inheritance Networks." UCLA, Cognitive Systems Laboratory, Technical Report 870053 (R-96).

Pearl, J. (1988) *Probabilistic Reasoning in Intelligent Systems: Networks of Plausible Inference.* Morgan Kaufmann, San Mateo, Calif.

Pearl, J. (1990a) "Jeffrey's Rule, Passage of Experience and Neo-Bayesianism." Pp. 121–35 in *Knowledge Representation and Defeasible Reasoning*, ed. H. Kyburg, R. Loui, & G. Carlson. Kluwer, Dordrecht.

Pearl, J. (1990b) "System Z: A Natural Ordering of Defaults with Tractable Applications to Default Reasoning." Pp. 121–35 in *Proceedings of TARK-90*, ed. R. Parikh. Morgan Kaufmann, San Mateo, Calif.

Pearl, J. (1993) "From Conditional Oughts to Qualitative Decision Theory." In D. Heckerman and A. Mamdani (eds.), *Proceedings of the Ninth Conference on Uncertainty in Artificial Intelligence (UAI-93)*, Washington, D.C., 12–20 July 1993. Morgan Kaufmann, San Mateo, Calif.

Pearl, J., & Verma, T. (1991) "A Theory of Inferred Causation." In J. A. Allen, R. Fikes, and E. Sandewall (eds.), *Principles of Knowledge Representation and Reasoning: Proceedings of the Second International Conference*, Cambridge, Mass., pp. 441–452, April 1991. Morgan Kaufmann, San Mateo, Calif.

Reichenbach, H. (1956) *The Direction of Time.* University of California Press, Berkeley.

Reiter, R. (1987) "Nonmonotonic Reasoning." *Annual Review of Computer Science* 2: 147–86.

Skyrms, B. (1981) "The Prior Propensity Account of Subjunctive Conditionals." Pp. 259–65 in *Ifs*, ed. W. L. Harper, R. Stalnaker, & G. Pearce. D. Reidel, Dordrecht.

Spohn, W. (1980) "Stochastic Independence, Causal Independence, and Shieldability." *Journal of Philosophical Logic* 9: 73–99.

Spohn, W. (1988a) "Ordinal Conditional Functions: A Dynamic Theory of Epistemic States." Pp. 105–34 in *Causation in Decision, Belief Change, and Statistics*, ed. W. L. Harper & B. Skyrms. D. Reidel, Dordrecht.

Spohn, W. (1988b) "A General Non-probabilistic Theory of Inductive Reasoning." Pp. 315–22 in *Proceedings of the Fourth Workshop on Uncertainty in Artificial Intelligence*, Minneapolis. Reprinted, 1990 (pp. 149–58), in *Uncertainty in Artificial Intelligence*, ed. R. D. Schacter, T. J. Levitt, L. N. Kanal, & J. F. Lemmer. North Holland, Amsterdam.

Stalnaker, R. (1981) "Letter to David Lewis." Pp. 151–2 in *Ifs*, ed. W. L. Harper, R. Stalnaker, & G. Pearce. D. Reidel, Dordrecht.

Suppes, P. (1970) *A Probabilistic Theory of Causation.* North Holland, Amsterdam.

6

The hypothesis of the conditional construal
of conditional probability

ALAN HÁJEK AND NED HALL

I'm sure that I could be a movie star
If I could get out of this place
—Billy Joel, "The Piano Man"

1. WHAT IS THE HYPOTHESIS?

"Conditional probability is the probability of a conditional." This statement, though catchy, leaves underspecified the exact content of the
hypothesis that we wish to discuss (hereafter, "the Hypothesis"). In fact, it
is not easy to characterize this content, because our concern is not with
some mathematical hypothesis (although, to be sure, it does have mathematical import), but rather with a hypothesis that is supposed to be able
to explain, among other things, certain features of ordinary language. We
begin with the equation that captures the leading idea behind the *conditional
construal of conditional probability* (CCCP):

$$P(A \rightarrow B) = P(B|A)$$

where P is a probability function, "$\rightarrow$" is a conditional connective, and
$P(B|A)$ is the conditional probability of B on A, given by the usual ratio
formula. There are three types of variables in this formulation: the
probability function, the "$\rightarrow$", and the propositions to appear on either
side of the "$\rightarrow$". We should specify appropriate quantifiers and domains
of quantification. Let us first make clear our quantification over propositions:

CCCP: $P(A \rightarrow B) = P(B|A)$ for all A, B in the domain of P, with $P(A) > 0$

Now let us quantify over the probability functions and the conditionals –
not in all the ways that one might conceive of, but rather so as to distinguish

75

four particularly important versions of the Hypothesis:

Universal version: There is some "→" such that for all P, CCCP holds.

Belief function version: There is some "→" such that for all P that could represent a rational agent's system of beliefs, CCCP holds.

Universal tailoring version: For each P there is some "→" such that CCCP holds.

Belief function tailoring version: For each P that could represent a rational agent's system of beliefs, there is some "→" such that CCCP holds.

The first version has been refuted a number of times, as we shall see; the third version has also been refuted. The second and fourth versions are, of course, still underspecified: Who is the agent? A rational human being? An ideally rational being? It will turn out that, to a great extent, it does not matter, because, as we shall show, the second version is untenable however it is construed, and we shall argue that even the fourth version is in a precarious state.

In what follows, when we talk about "the Hypothesis," often we shall leave open exactly which version we mean, except when it matters.

2. WHY CARE ABOUT THE HYPOTHESIS?

The idea that the probability of a conditional equals the corresponding conditional probability goes back at least to Jeffrey (1964), who thought that it could illuminate a theory of confirmation, and Ellis (1969), who assumed it when arguing that truth logic is a special case of probability logic. In a sense, it finds a precursor even as far back as de Finetti (1936), as we shall see later. However, its best-known presentation is due to Stalnaker (1970). Thus, it came to be known as "Stalnaker's Hypothesis," at once an unhappy and a happy name – unhappy, because Stalnaker himself no longer holds the Hypothesis, and indeed has provided cogent arguments against it, which we shall discuss; happy, because the Hypothesis has been a rich source of philosophical debate, and Stalnaker should take much of the credit for this.

Nowadays it is more Adams' name that we should associate with it, because a certain variant of it – the so-called Adams Thesis, which we shall also discuss – has much currency, thanks to his work on conditionals. This, too, is both a happy and an unhappy name – happy, because of the ongoing research the thesis has prompted (e.g., see the contribution to this volume by Stalnaker and Jeffrey, Chapter 4); unhappy, because Adams no longer holds the thesis as it is usually stated. Adams' Thesis as it is usually stated involves the assertability of the indicative conditional (e.g., Jackson, 1987). Adams invited this interpretation in his 1965 paper, where he made frequent use of the term "assertability." However, this talk of assertability had disappeared from his writings by the time of his 1975 book, to be replaced by a notion that we shall call "quasi probability." Adams has informed us (pers. commun.) that what he had in mind had more to do with reasonableness of belief than with a notion of appropriateness of utterance suggested by the term "assertability." We use the qualification "quasi" because, as will be noted later, Adams' quasi-probabilities do not conform to the usual probability calculus.

Stalnaker (1970) and Adams (1965, 1975) had similar motivations: They intended the Hypothesis to illuminate the semantics of the conditional. Stalnaker's idea was roughly that the Hypothesis would serve as a criterion of adequacy for a truth-conditional account of the conditional. Stalnaker had already provided such an account (1968), but Lewis was working at the same time on his account (which was to appear in 1971 and to be more fully elaborated in 1973), and – as Stalnaker knew – the two accounts disagreed. The Hypothesis provided Stalnaker a forceful argument for his side of the debate, because, as we shall see, it entails (a version of) the conditional excluded middle, and acceptance of this is the main point of disagreement between the two accounts. In the 1970 paper, Stalnaker's strategy is this: Rather than looking directly for the truth conditions of the conditional, he considers its belief conditions – that is, conditions under which it would be reasonable for a rational agent to believe a given conditional. He models the belief system of a rational agent as an assignment P of subjective probabilities to worlds in some probability space; the agent believes a conditional $A \rightarrow B$ exactly when his or her subjective probability for it is sufficiently high. Now the Hypothesis enters, identifying the subjective probability for $A \rightarrow B$ with $P(B|A)$, which in turn is straightforwardly determined by the agent's assignment of probabilities to the propositions A and A & B, provided that $P(A) > 0$.[1] So the agent believes the conditional just in case $P(B|A)$ is sufficiently high. Although such belief conditions for the "$\rightarrow$" do not fix its truth conditions, they do put constraints on them – sufficient to rule against the Lewis account, for example.

Throughout a good part of the 1960s and 1970s, Adams was engaged in the project of supplementing the traditional truth-conditional notion of validity of arguments with a "probabilistic soundness" criterion of his own. It finds its fullest treatment in Adams' 1975 book, on which we shall focus our discussion. Adams believed that conditionals do not have truth conditions, and thus he found the traditional approach inadequate for arguments in which conditionals appear; but he was happy to speak of "probabilities" attaching to conditionals. Roughly, a probabilistically sound argument is one for which it is impossible for the premises to be probable while the conclusion is improbable. He invoked the Hypothesis to govern the assignment of probabilities to indicative conditionals and argued that the resulting scheme respected intuitions about which of the inferences were reasonable, and which not. Although Adams appealed to the same Hypothesis (at least superficially) as Stalnaker, unlike Stalnaker he did not regard the probability of a conditional as the probability of its *truth*. Further, Adams' "probabilities" of conditionals do not conform to the usual probability calculus – thus the suggestion by Lewis (1976) that they be called "assertabilities" instead, a practice that has been widely adopted subsequently. Given that Adams never explicitly disavowed such talk, this suggestion was certainly reasonable; recall, however, that it conflicts with Adams' own views, as noted earlier.

Thus it seems that Stalnaker proposed the belief function version of the Hypothesis and that Adams proposed a near variant of it. Call a probability function P that conforms to CCCP for a given choice of "$\rightarrow$" *a CCCP-function for* "$\rightarrow$". Stalnaker's original proposal, then, was that there exists an "$\rightarrow$" such that (1) "$\rightarrow$" can plausibly be construed as a conditional, and (2) the set of CCCP-functions for this "$\rightarrow$" includes all probability functions that could represent a rational agent's system of beliefs (for short: all *belief functions*). Indeed, he believed that the Stalnaker conditional was just such an "$\rightarrow$".

This proposal has since been dashed by an impressive battery of results that count against the truth of the Hypothesis – although not decisively against all its versions. We shall provide a survey of these results here. We shall also consider in Section 7 the extent to which Adams' variant of the Hypothesis is immune to them.

The situation is interesting, whatever the significance of the results turns out to be. If the Hypothesis (on any of its versions) is false, then seemingly synonymous locutions are not in fact synonymous: Surprisingly, "the probability of B, given A" does not mean the same thing as "the probability of: B if A." If the Hypothesis (on any of its versions) is true, then it establishes important links between logic and probability theory, as Stalnaker and Adams hoped it would. As Stalnaker (1970) observed, "although the interpretation of probability is controversial, the abstract calculus is a relatively well defined and well established mathematical theory. In contrast to this, there is little agreement about the logic of conditional sentences. ... Probability theory could be a source of insight into the formal structure of conditional sentences" (p. 107).[2] For example, the material conditional does not conform to CCCP, as we shall see (at least when the P in the equation is genuine probability, rather than assertability); this in turn legislates against the Hypothesis applying to the indicative conditional, according to one major account of that conditional [favored

78

by Jackson (1987) and Lewis (1976), among others]. Or again, the Hypothesis entails (a probabilistic version of) the conditional excluded middle; this in turn legislates against Lewis' counterfactual logic, and in favor of Stalnaker's.

And the Hypothesis could similarly enrich probability theory, because it would assist in the interpretation of conditional probability. Recall de Finetti's (1972) lament that the usual ratio gives the formula for, but not the meaning of, conditional probability. The Hypothesis could serve to characterize more fully what the ratio means, and what its use is.

Finally, the Hypothesis would solve what van Fraassen (1989) calls "the Judy Benjamin problem," a problem in probability kinematics. The general problem for probability kinematics is this: Given a prior probability function P, and the imposition of some constraint on the posterior probability function, what should this posterior be? This problem has a unique solution for certain constraints. For example:

1. Assign probability 1 to some proposition E (while preserving the odds of all propositions that imply E). (Solution: Conditionalize P on E.)
2. Assign probabilities $p_1, \ldots, p_n$ to the cells of the partition $\{E_1, \ldots, E_n\}$ (while preserving the odds of all propositions within each cell). (Solution: Jeffrey-conditionalize P on this partition, according to the specification.[3])

But consider this constraint:

3. Assign conditional probability p to B, given A.

The Judy Benjamin problem is that of finding a rule for transforming a prior, subject to this third constraint.

Van Fraassen provides arguments for three distinct such rules, and he surmises that this raises the possibility that uniqueness results "will not extend to more broadly applicable rules in general probability kinematics. In that case rationality will not dictate epistemic procedure even when we decide that it shall be rule governed" (1989, p. 343). But if the Hypothesis were true, a particularly simple solution would present itself. After all, constraint 3 would then be equivalent to the following:

3'. Assign probability p to $A \rightarrow B$.

This is uniquely met by a simple Jeffrey conditioning, on the partition $\{A \rightarrow B, \neg(A \rightarrow B)\}$ (assuming that the odds of propositions within each cell are to remain the same).

One conclusion of this chapter will be that the first three versions of the Hypothesis are not tenable. Nevertheless, it is noteworthy that it is surprisingly difficult to produce an intuitive counterexample to CCCP.

79

This suggests that we should consider the fourth version, weak though it is. And yet even this can be held only at considerable cost.

3. WHY BELIEVE THE HYPOTHESIS?

Before arguing against the Hypothesis, we want to make the case in favor of it as strong as possible. The following sections offer some reasons for believing it.

It sounds right

Both sides of CCCP seem to "say" the same thing. As van Fraassen (1976) wrote, "the English statement of a conditional probability sounds exactly like that of the probability of a conditional. What is the probability that I throw a six if I throw an even number, if not the probability that: if I throw an even number, it will be a six?" (pp. 272–3). Many assertions of conditional probability sound like assertions *of probability*, within the scope of which is something that seems best analyzed as a conditional. Thus, the Hypothesis certainly does justice to the grammar of conditional probability statements. And case-by-case evidence, such as van Fraassen's example here, seems to support it. This surely explains the initial appeal of the Hypothesis.

Ramsey's test

Ramsey (1965) suggested that we evaluate the conditional "if A, then B" as follows: First, we hypothetically add A to our system of beliefs, minimally revising what we currently believe in order to do so; second, we evaluate B on the basis of our revised body of beliefs. $P(A \to B)$ measures how well the conditional performs on Ramsey's test. But apparently $P(B|A)$ does too, for conditioning on A prima facie seems to capture the notion of "minimally revising what we currently believe in order to accommodate A"; and our evaluation of B in our new belief state is just $P(B|A)$.

Adams' Thesis[4]

Assertability is said to go by subjective probability of truth. [Or at least it usually does, though not always: For example, the assertability of "A but B" differs from that of "A and B"; yet the two have the same subjective probability, namely, $P(A \ \& \ B)$.] At first sight, the indicative conditional appears to provide a counterexample to this dictum. After all, according to Adams' Thesis, the assertability of $A \to B$ equals $P(B|A)$ – and even

80

someone who, unlike Adams, believes that the indicative conditional has truth conditions may find Adams' Thesis about its assertability compelling. (Of course, someone *like* Adams thinks that the indicative conditional provides a counterexample to the dictum, because according to him, there is no such thing as the probability of its *truth*.) It would be nice for proponents of the truth-conditional view of the indicative conditional if it were no exception to the dictum. If the Hypothesis were true, it would then explain Adams' Thesis admirably: The assertability of $A \to B$ goes by $P(A \to B)$ (as per the dictum), which in turn equals $P(B|A)$.

Stalnaker validity

Stalnaker (1970) deployed the Hypothesis in support of his C2 logic for the conditional by proving the coincidence of two sorts of validity: The sentences that are valid under his truth conditions for the conditional turn out to be exactly those that receive conditional probability of 1, given every condition, assuming the truth of CCCP.

Adams' probabilistic soundness

As we indicated earlier, Adams found that his variant of the Hypothesis, coupled with his probabilistic soundness criterion for arguments, gives verdicts that accord with our intuitions on the acceptability of arguments. In particular, it classifies as fallacious the various notorious "paradoxes of material implication," such as the inference from "not-A" to "if A, then B."

Independence of the conditional from its antecedent

CCCP is equivalent to A being probabilistically independent of $A \to B$ according to P, for all A and B, under the usual assumption of *centering*.[5] This assumption, common to Stalnaker's and Lewis' logics, states that $(A \to B) \& A$ is equivalent to $A \& B$. With CCCP and $P(A) > 0$, it follows that $P[A \& (A \to B)] = P(A \& B) = P(A)P(B|A) = P(A)P(A \to B)$; thus $P[A \& (A \to B)] = P(A)P(A \to B)$ for all A and B.[6] The reverse implication – that this independence condition together with centering entails CCCP – is obvious.

So given the assumption about the logic, CCCP hinges on whether or not $P[A \& (A \to B)] = P(A)P(A \to B)$ for all A and B. Furthermore, it might seem that a conditional *is* always independent of its antecedent. Prima facie, this is plausible at least for the counterfactual conditional. After all,

81

whether or not A is actually true might seem to be quite irrelevant to what *would be* the case *if A* were true.

The foregoing considerations provide a strong case in favor of (some version of) the Hypothesis. Unfortunately, the case against it is far more powerful. Indeed, problems arise even before we reach the famed "triviality results." As a prelude to discussion of those results and other technical material of interest, we consider some of these problems in the next section.

4. Sources of Suspicion

Here we canvass some reasons to be suspicious about the Hypothesis. They fall short of precise refutations, but set the stage for them.

The Hypothesis fails for the material conditional

The probability of the material conditional, $A \supset B$, is given by $P(A \supset B) = P(\neg A) + P(A \& B)$. It is easy to verify that if $1 > P(A) > 0$, $P(A \supset B) = P(B \mid A)$ iff $P(A) = P(A \& B)$; this condition clearly cannot obtain in general. Of course, this failure need not alarm a proponent of the Hypothesis who takes the conditional appearing in it to be something other than the material conditional. However, as we have noted, the Hypothesis clearly cannot apply to the indicative conditional, on one major account of that conditional. That, in turn, undermines to some extent one of the reasons for believing the Hypothesis: the nice explanation that it would provide for Adams' Thesis about the indicative conditional.

Independence of the conditional from its antecedent again: causal decision theory

Stalnaker's own suspicions about the Hypothesis arose from considerations of situations in which an outcome, or act, is evidentially relevant to, or *stochastically dependent on*, another outcome without *causally influencing* that outcome. Stalnaker (1976) drew this distinction as follows:

> A is stochastically dependent on B iff $P(B \mid A) \neq P(B)$, whereas
> A is causally independent of B iff $P(A \rightarrow B) = P(B)$.

If there really are cases such as Stalnaker imagines, then we immediately have the failure of CCCP, at least for the "$\rightarrow$" in terms of which causal independence is defined.[7] The conviction that there are such cases – Newcomb's problem, for instance – led to the development of various causal decision theories. Of course, these are cases of conditionals that are *not* independent of their antecedents, contrary to the intuition expressed in the preceding section. For instance, the truth value of "I choose both boxes → I get just the $1,000" in Newcomb's problem is thought to depend on whether or not I choose both boxes.

Centering

The Hypothesis, together with two very weak assumptions about the logic of the "→," guarantees a probabilistic version of a much stronger such assumption. The principles are *modus ponens* – $(A \rightarrow B)\& A$ entails $A\&B$ – and *entailment within the consequent* – $(A \rightarrow B)\&(A \rightarrow C)$ entails $A \rightarrow (B\&C)$.[8] With CCCP, these yield[9] $P[A\&(A \rightarrow B)] = P(A\&B)$ for all A and B, a probabilistic version of the logical principle of centering (which, recall, states that $(A\&B) = (A\&A \rightarrow B)$ for all A and B).

Those uncomfortable with the logical version should also balk at this probabilistic version. Here is one reason for being suspicious about both versions, at least for counterfactuals. Consider some indeterministic process that is certain to occur – as it might be, a particular toss of a fair and genuinely indeterministic coin. It seems reasonable to claim that the probability of the "would" counterfactual "if the coin were tossed, it would land heads" is *zero*, because the contradictory "might" counterfactual seems certainly true: "if the coin were tossed, it might not land heads." But the probabilistic version of centering requires the probability of the "would" counterfactual to be at least as great as the probability of the corresponding conjunction. And because the coin is certain to be tossed, that probability is surely $\frac{1}{2}$. If this argument is right, it follows – on pain of losing the very weak and plausible assumptions about the logic of the "→" – that the Hypothesis does not hold for the counterfactual conditional.

Conditional excluded middle

The distinctive feature of Stalnaker's logic for conditionals is its adoption of the conditional excluded middle (CEM): For any A and B, it is an axiom that $(A \rightarrow B) \vee (A \rightarrow \neg B)$. Although it is still controversial whether this feature is a virtue or a vice, CCCP, together with just one of the aforementioned weak logical principles governing the "→", implies the following probabilistic version of it: For all A and B such that $P(A) > 0$, $P[(A \rightarrow B) \vee (A \rightarrow \neg B)] = 1$ (we leave the demonstration for the next section). So putative counterexamples to CEM will serve as putative counterexamples to CCCP. Consider, for instance, Lewis' (1973, p. 80) critique of CEM, using Quine's (1950, p. 14) famous example, with A taken to be "Bizet and Verdi are compatriots," and B taken to be "Bizet and Verdi are Italian." On Lewis' view, at least one of $P(A \rightarrow B)$ and $P(A \rightarrow \neg B)$

must be less than CCCP predicts. Similarly, indeterminism may cast doubt on CEM (although intuitions vary here). Suppose again that coin tosses are genuinely indeterministic. Then both the conditionals "if the coin were tossed, it would land heads" and "if the coin were tossed, it would not land heads" appear to be false.

These bad omens give way to sharp arguments in the triviality results discussed in the next section.

5. TECHNICAL RESULTS

5.1. Introduction

Since its debut, the Hypothesis has given rise to a wealth of technical work, the most significant elements of which we shall review here. Most of these results have been negative in focus, designed to show that the Hypothesis contradicts various purportedly plausible hypotheses. But some results – notably those van Fraassen published in his 1976 paper – have demonstrated that, under certain conditions, the Hypothesis can be maintained. Whether or not these conditions are so stringent as to make the Hypothesis philosophically uninteresting is a good question, which we shall address in the next section. For now, our aim is only to summarize the most significant results as precisely and succinctly as possible.

Before proceeding, however, we wish to emphasize that the "slogan" form of the Hypothesis with which we began may give rise to the misconception that in order to refute the Hypothesis, it suffices to exhibit, for any "$\rightarrow$," a single probability function and a pair of propositions in its domain for which the equation fails. To be sure, that refutes the universal version; but doing *that* is easy, and it leaves the weaker versions untouched. To see just how easy it is, note that (apart from special cases) the family of CCCP-functions for a given "$\rightarrow$" is not closed under mixing.[10] For example, suppose that P_1 and P_2 are CCCP-functions for a given "$\rightarrow$," and let $P_3 = \frac{1}{2}P_1 + \frac{1}{2}P_2$. Then it is an easy exercise (which we leave to the reader) to verify that P_3 will not, in general, be a CCCP-function for this "$\rightarrow$." Needless to say, the results we shall canvass all aim at much greater generality than this.

5.2. Terminology

Some preliminary definitions will facilitate the classification of results. We introduce the term *probability space* (abbreviation: "space") to denote a triple $\langle W, F, P \rangle$, and the term *model* to denote a quadruple $\langle W, F, P, \rightarrow \rangle$;

84

we shall call $\langle W, F, \rightarrow \rangle$ a *conditional algebra* (abbreviation: "algebra"). W is a set (heuristically: a set of possible worlds); F is a sigma-field of subsets of W (that is, it is closed under the Boolean operations, including countable union) (heuristically: a set of significant propositions); P is a countably additive probability measure on F; and "$\rightarrow$" is a binary operator on F, such that for all $A, B \in F, A \rightarrow B \in F$.

We say that *CCCP holds for* a model $\langle W, F, P, \rightarrow \rangle$ iff for all $A, B \in F$, where $P(A) > 0$, $P(A \rightarrow B) = P(B \mid A)$. In that case, "$\rightarrow$" is a *CCCP-conditional* for the space $\langle W, F, P \rangle$, and P is a *CCCP-function* for the algebra $\langle W, F, \rightarrow \rangle$.

Call a model $\langle W, F, P, \rightarrow \rangle$ or space $\langle W, F, P \rangle$ *trivial* iff P has at most four conditional probability values.[11] It is easy to construct an "$\rightarrow$" that will extend a trivial probability space $\langle W, F, P \rangle$ to a trivial model for which CCCP holds.[12] Hence the typical conclusion of the negative results, that the Hypothesis holds for a range of models only if they are all trivial.

Given a prior specification of a set of worlds W and a set of significant propositions F, we can restate the four versions of the Hypothesis introduced earlier:

Universal version: There is some "$\rightarrow$" such that CCCP holds for all models $\langle W, F, P, \rightarrow \rangle$.

Belief function version: There is some "$\rightarrow$" such that CCCP holds for all models $\langle W, F, P, \rightarrow \rangle$ where P could represent a rational agent's system of beliefs.

Universal tailoring version: For each P there is some "$\rightarrow$" such that CCCP holds for the model $\langle W, F, P, \rightarrow \rangle$.

Belief function tailoring version: For each P that could represent a rational agent's system of beliefs, there is some "$\rightarrow$" such that CCCP holds for the model $\langle W, F, P, \rightarrow \rangle$.

The results we shall discuss can all be seen as addressing, with the aid of auxiliary assumptions, one or more of these versions. Before turning to them, we shall consider briefly what constraints we might impose on a model $\langle W, F, P, \rightarrow \rangle$ by virtue of the fact that "$\rightarrow$" is supposed to be a *conditional*. It turns out that a fourfold hierarchy is particularly useful:

First, we might impose no constraints at all: for any acceptable model $\langle W, F, P, \rightarrow \rangle$, "$\rightarrow$" need only be a binary operator on F.

Second, we might require that *modus ponens* hold. In other words, any acceptable model $\langle W, F, P, \rightarrow \rangle$ most obey the following principle:[13]

(L1) For all $A, B \in F, A \cap (A \rightarrow B) \subseteq AB$.

Third, we might require in addition that *entailment within the consequent* hold. In other words, any acceptable model $\langle W, F, P, \rightarrow \rangle$ must obey the following principle:

(L2) For all $A, B, C \in F, (A \to B) \cap (A \to C) \subseteq A \to (BC)$.

Finally, we might require in addition that *weakened transitivity* hold. In other words, any acceptable model $\langle W, F, P \to \rangle$ must obey the following principle:

(L3) For all $A, B, C \in F, (A \to B) \cap (B \to A) \cap (B \to C) \subseteq A \to C$.

It may seem surprising that we have included in our logical hierarchy only three principles that might govern the "$\to$." In fact, no other principles are needed; wherever a further principle seems necessary, a probabilistic equivalent will do, and that equivalent is always readily derivable from the given together with CCCP. For example, centering is often supposed necessary to derive the independence of the conditional from its antecedent from CCCP (see Sections 3 and 4). But a quick scan of the proof reveals that all that is needed is that $P(AB) = P(A \cap A \to B)$ – and *this*, as will shortly be seen, follows easily from CCCP together with (L1) and (L2).[14] We seek in the next section to clarify matters by detailing a number of consequences of these two principles and CCCP. Because these consequences concern probabilistic equivalents of what might otherwise have been introduced as logical axioms, we call them "probabilistic entailment results."

5.3. Probabilistic entailment results

In what follows, we shall, with harmless ambiguity, sometimes denote the complement of a proposition A as $\bar{A}$ and sometimes as $\neg A$. Then for any model $\langle W, F, P, \to \rangle$ that obeys (L1), (L2), and CCCP, all of the following hold:

(PE1) For all $A, B \in F$, if $A \subseteq B$ and $P(A) > 0$, then $P(A \to B) = 1$.

(PE2) For all $A, B \in F$, if $P(AB) = 0$ and $P(A) > 0$, then $P(A \to B) = 0$.

(PE3) For all $A, B, C \in F$ such that $P(BC) = 0$, if $P(A) > 0$, then $P[(A \to B) \cap (A \to C)] = 0$.

(PE4) For all $A, B \in F$, if $P(A) > 0$, then $P[(A \to B) \cup (A \to \bar{B})] = 1$.

(PE5) For all $A, B, C \in F$, if $P(A) > 0$, then $P[C \cap \neg (A \to B)] = P[C \cap (A \to \bar{B})]$.

(PE6) For all $A, B, C \in F$, if $P(A) > 0$, then $P(A \to B \cap A \to C) = P(A \to BC)$.

(PE7) For all $A, B, C \in F$, if $P(A) > 0$, then $P(A \to B \cup A \to C) = P[A \to (B \cup C)]$.

(PE8) For all $A, B \in F$, if $P(A \cup B) > 0$, then $P[((A \cup B) \to A) \cup ((A \cup B) \to B)] = 1$.

(PE9) For all $A, B \in F, P(AB) = P(A \cap A \to B)$.

(PE1) and (PE2) are obvious, and follow from CCCP alone. Adding principle (L2), we have, for $P(A) > 0$, $P[(A \to B) \cap (A \to C)] \leqslant P(A \to BC) = P(BC|A)$, from which (PE3) follows at once. Therefore, $P[(A \to B) \cup (A \to \bar{B})] = P(A \to B) + P(A \to \bar{B}) - P[(A \to B) \cap (A \to \bar{B})] = P(B|A) + P(\bar{B}|A) = 1$, yielding (PE4). The proof of (PE5) is then straightforward:

$$P[C \cap \neg (A \to B)] = P\{[C \cap \neg (A \to B)] \cap [(A \to B) \cup (A \to \bar{B})]\}$$
$$[\text{by (PE4)}]$$
$$= P[C \cap \neg (A \to B) \cap (A \to \bar{B})]$$
$$= P(C \cap A \to \bar{B}) - P(C \cap A \to \bar{B} \cap A \to B)$$
$$= P(C \cap A \to \bar{B}) \qquad [\text{by (PE3)}]$$

(PE5), (PE3), and (L2) thus guarantee that, for $P(A) > 0$, $P(A \to B \cap A \to C \cap \neg (A \to BC)) = 0$. Thus, $P(A \to B \cap A \to C) = P(A \to B \cap A \to C \cap A \to BC) = P(A \to BC) - [P(A \to \bar{B} \cap A \to BC) + P(A \to B \cap A \to \bar{C} \cap A \to BC)] = P(A \to BC)$, again making repeated use of (PE3) and (PE5); this establishes (PE6). (PE7) is then almost immediate: $P(A \to B \cup A \to C) = P(A \to B) + P(A \to C) - P(A \to B \cap A \to C) = P(A \to B) + P(A \to C) - P(A \to BC) = P(B|A) + P(C|A) - P(BC|A) = P(B \cup C|A) = P[A \to (B \cup C)]$. Of course, (PE8) follows at once from (PE7) and CCCP. Finally, we add (L1) to obtain (PE9), as follows: If $P(A) = 0$, the claim follows trivially, so suppose otherwise. Then $P(AB) = P(AB \cap A \to B) + P(AB \cap A \to \bar{B}) = P(AB \cap A \to B) = P(A \cap A \to B) - P(A\bar{B} \cap A \to B) = P(A \cap A \to B)$.

Thus, we see that from CCCP, together with (*L1*) *and* (*L2*) *alone*, follow probabilistic equivalents of every principle that is characteristic of Stalnaker's logic C2, save only the weakened transitivity condition (L3). And probabilistic equivalents are enough for the triviality results, to which we shall now turn.

5.4. *Selective overview of triviality results*

We shall say that two models $\langle W_1, F_1, P_1, \to_1 \rangle$ and $\langle W_2, F_2, P_2, \to_2 \rangle$ *employ the same arrow* iff the corresponding algebras $\langle W_1, F_1, \to_1 \rangle$ and $\langle W_2, F_2, \to_2 \rangle$ are isomorphic. (We shall also sometimes say that in such a case, the two models employ a *uniform interpretation of the conditional*.) We can thus sensibly speak of *the* algebra (up to isomorphism) that is associated with a set of models that employ the same arrow, and then go on to ask which probability functions definable on this algebra the set includes, consistent with CCCP holding nontrivially for all models in the set. That is, given an algebra, what are its nontrivial CCCP-functions? With respect to this question, triviality results come in two flavors. The first kind, which we shall call *no-go results*, demonstrate that there are *no* nontrivial CCCP-functions for any algebra with such-and-such features. The second kind, which we shall call *limitation results*, demonstrate that the class of nontrivial CCCP-functions for any algebra with such-and-such features is severely limited in some respect. The distinction is somewhat arbitrary, but we trust that its usefulness will be apparent in what follows. We shall review limitation results first.

As is customary, we shall say that a probability function P_C is derived from P by *conditioning* if there is some proposition C such that for all $X, P_C(X) = P(X|C)$. Similarly, we shall say that P_x is derived from P by *nondegenerate two-celled Jeffrey conditioning* if there is a proposition C, and an x with $0 < x < P(\neg C)$, such that for all $B, P_x(B) = P(B) + x[P(B|C) - P(B|\neg C)]$.[15] We shall say that a class of probability functions is *closed under conditioning* (*Jeffrey conditioning*) iff any probability function derived from a function in the class by conditioning (Jeffrey conditioning) is itself in the class.

Finally, a proposition $A \in F$ is a *P-atom* iff $P(A) > 0$ and, for all X, either $P(AX) = 0$ or $P(AX) = P(A)$.

Lewis' triviality results. David Lewis has four closely related limitation results, the first two published in 1976, and the third and fourth in 1986. They make no assumptions about the logic of the "→." Given any algebra $\langle W, F, \rightarrow \rangle$, Lewis' results reveal that the set S of CCCP-functions for this algebra is limited in the following respects:

First triviality result. S does not contain all probability functions definable on the given algebra.

Second triviality result. Unless it contains only trivial functions, S is not closed under conditioning.

Third triviality result. Unless it contains only trivial functions, S is not closed under conditioning restricted to the propositions in a single finite partition.

Fourth triviality result. Unless it contains only trivial functions, S is not closed under nondegenerate two-celled Jeffrey conditioning.

Lewis' results deserve careful scrutiny, particularly because they are so well known and widely cited. To begin, note that the third result implies the second, which in turn implies the first; so from a logical point of view there are really only two results here.[16] More importantly, it turns out that using an idea from Lewis, we can derive an even stronger result, which has his first three triviality results as corollaries.

The key maneuver in Lewis' proofs is to demonstrate the following: If $\langle W, F, P, \to \rangle$ and $\langle W, F, P_C, \to \rangle$ both obey CCCP (and employ the same "$\to$"), and P_C is derived from P by conditioning on C, then for all $A, B \in F$ such that $P(AC) \neq 0$, $P(A \to B | C) = P(B | AC)$. This is easy to show: $P(A \to B | C) = P_C(A \to B) = P_C(B | A) = P(B | AC)$. The equality of the second and third terms requires the uniform interpretation of the conditional; for suppose that the interpretations differed, so that the second CCCP-model was not $\langle W, F, P_C, \to \rangle$ but rather $\langle W, F, P_C, \to^* \rangle$, where "$\to^*$" $\neq$ "$\to$." Then we would have $P(A \to B | C) = P_C(A \to B)$ and $P_C(A \to^* B) = P_C(B | A) = P(B | AC)$, but not necessarily the needed $P_C(A \to B) = P_C(A \to^* B)$. We shall now show that uniformity of the conditional over just *these two* models is sufficient to derive triviality (and of course this falls far short of uniformity over an entire class of models whose probability functions are closed under conditioning, or restricted conditioning).

Theorem (the "Strengthened Lewis Result"). If $\langle W, F, P, \to \rangle$ and $\langle W, F, P_C, \to \rangle$ are distinct nontrivial models, with P_C derived from P by conditioning on C, then CCCP fails for at least one of them.

Proof. Suppose for *reductio* that $\langle W, F, P, \to \rangle$ and $\langle W, F, P_C, \to \rangle$ are distinct and nontrivial and that CCCP holds for both. Their distinctness implies that $P(C) < 1$. Nontriviality of the latter guarantees that C is not a P-atom – for conditioning on a P-atom yields a probability function with just the values 0 and 1. Because C is not a P-atom, there is some $D \subset C$ such that $0 < P(D) < P(C)$. Choose such a D. Let $E = D \cup \bar{C}$. Then $P(E) < 1$. But

$$P(E \to \bar{C}) = P(E \to \bar{C} | C)P(C) + P(\bar{C} \cap E \to \bar{C})$$
$$= P(\bar{C} | EC)P(C) + P(\bar{C} \cap E \to \bar{C})$$

$$\text{(by Lewis' key maneuver)}$$

$$= 0 + P(\bar{C} \cap E \to \bar{C})$$
$$\leqslant P(\bar{C}).$$

Applying CCCP to the left-hand side and rearranging yields $P(\bar{C}) \leqslant P(E)P(\bar{C})$, whence $P(E) \geqslant 1$ – a contradiction that completes the *reductio*.

Thus, the technique Lewis employs to prove the first three triviality results can be easily adapted to establish something more telling: In the set S of nontrivial CCCP-functions for a given algebra, no function is the conditionalization of any other. The fourth triviality result is, of course, logically independent of the Strengthened Lewis Result. However, both have now been superseded by much stronger results.

Hall's first triviality result. Hall's Chapter 8 in this volume contains three new results pertaining to CCCP, the first of which we cover here. Hall shows that, given no assumptions about the logic of the "→," the set S of nontrivial CCCP-functions for a given algebra is limited as follows:

Orthogonality Result. For any $P, P' \in S$, if $P \neq P'$, then P and P' are *orthogonal*; that is, for some $A, P(A) = 1$ but $P'(A) = 0$.

Because no two functions related by conditioning are orthogonal, and no two functions related by nondegenerate Jeffrey conditioning (two-celled or otherwise) are orthogonal, Lewis' four results – as well as the Strengthened Lewis Result – follow immediately. Indeed, if we conceive of conditioning and Jeffrey conditioning as candidates for *updating rules* – that is, rules specifying how probabilities should be modified under the impact of new evidence – then, because it is plausible that no two functions related by a reasonable updating rule will be orthogonal, we can make this stronger claim: For any algebra and any CCCP-function P for that algebra, no function derived from P via some updating rule will be a CCCP-function for the given algebra.

Hájek's second triviality result. Hájek, in Chapter 7 of this volume, provides a recipe for generating counterexamples to CCCP, based on the notion of a *perturbation*. Intuitively (and for less intuition and more precision, the reader is referred to Hájek's Chapter 7), a perturbation of a function P is a distinct function P' that agrees with P in some specified way, but disagrees with it in some other specified way. Developing the notion of perturbation in more detail, Hájek demonstrates the following limitation on the set S of nontrivial CCCP-functions for a given algebra:

Perturbation Result. No member of S is the perturbation of any other.

Hájek shows, *inter alia*, that three of Lewis' four triviality results follow from the perturbation result (modulo a slightly different understanding of the word "trivial"). Now, given Hájek's definition of the term, no perturbation of a function is orthogonal to that function. So the Perturbation Result, as stated, follows from the Orthogonality Result. However, Hájek in fact shows something more specific: If P conforms to CCCP for a particular choice of propositions A and B – that is, if $P(A \to B) = P(B \mid A)$ – then any perturbation of P fails to conform to CCCP for *that very choice* of A and B. This fact will be relevant later, when we consider an attempt to save the Hypothesis by restricting the domain of propositions for which CCCP is supposed to hold.

This completes the exposition of the most important limitation results. We turn now to the no-go results.

Hájek's first triviality result. Making no assumptions about the logic of the "→," Hájek (1993) shows that for no nontrivial model $\langle W, F, P, \rightarrow \rangle$ that obeys CCCP is it the case that P takes on only a finite range of values. The following corollary (first proved in his 1989 paper) is immediate:

Finitude Result. The set S of nontrivial CCCP-functions for an algebra $\langle W, F, \rightarrow \rangle$ is empty if W is finite.

The central idea behind Hájek's proof is quite simple. Hájek observes that for any probability function P, the number of distinct values of $P(\cdot \rightarrow \cdot)$ is no greater than the number of unconditional probability values of P, and that in turn is no greater than the number of *conditional* probability values of P – for any unconditional value $P(A)$ is just the same as the conditional value $P(A | W)$. The bulk of the proof then consists in establishing that, for functions with a finite range, the number of conditional values is *strictly greater* than the number of unconditional values. The upshot is that for each nontrivial finite model, there is some conditional probability value that equals none of the unconditional probability values, and so *a fortiori* equals none of the $P(\cdot \rightarrow \cdot)$ values.

Hall's second and third triviality results. By assuming just that the conditional has enough logical structure to obey (L1) (*modus ponens*), Hall (Chapter 8) is able to demonstrate the following:

No Atoms Result. No nontrivial model $\langle W, F, P, \rightarrow \rangle$ that obeys CCCP and (L1) contains a P-atom.

Recall that a proposition A is a P-atom iff $P(A) > 0$ and, for all X, either $P(AX) = P(A)$ or $P(AX) = 0$. It follows from the No Atoms Result that any nontrivial model that obeys CCCP and (L1) is *full*, where a model (or probability space) is full iff for each A and $r \in (0, P(A))$ there is a $B \subset A$ such that $P(B) = r$.[17] And from this, in turn, it follows that if $\langle W, F, P, \rightarrow \rangle$ obeys CCCP and (L1), then W is non-denumerable.

If we add the principle (L2), a more surprising result is available – although it is somewhat tedious to state. Suppose we have an artificial language whose sentences Φ, Ψ, and so forth, are composed of atomic sentences A, B, C, and so forth, together with the usual truth-functional connectives and the special two-place connective "→." We can then define an *interpretation* for this language to be a pair $\langle I, M \rangle$, where M is a model $\langle W, F, P, \rightarrow \rangle$ and I is a function assigning an element of F to each atomic sentence, and behaving in the usual way with respect to the connectives [in particular, $I(\Phi \rightarrow \Psi) = I(\Phi) \rightarrow I(\Psi)$]. Then we can assign probabilities to sentences relative to a given interpretation $\langle I, M \rangle$: $P(\Phi) = P(I(\Phi))$, where $P \in M$. Let us require that for any *allowable* interpretation

91

$\langle I, M \rangle$, M be nontrivial and obey CCCP, (L1), and (L2); let us further require that I assign to the atomic sentences $\mathbf{A}$, $\mathbf{B}$, and $\mathbf{C}$ disjoint propositions, each of which has positive probability.[18] Then we have the following:

Constructibility Result. There is an effective procedure for specifying, for any positive integer n, a set of sentences $\{\mathbf{\Phi}_1, \ldots, \mathbf{\Phi}_n\}$ such that the only atomic sentences appearing in $\mathbf{\Phi}_1, \ldots, \mathbf{\Phi}_n$ are $\mathbf{A}$, $\mathbf{B}$, and $\mathbf{C}$, and such that, relative to *any* allowable interpretation, (1) for any i, $P(\mathbf{\Phi}_i) = 1/n$, and (2) $P(\mathbf{\Phi}_i \& \mathbf{\Phi}_j) = 0$ if $i \neq j$.

It follows that there is a procedure for constructing, for any rational number $r \in [0, 1]$, a sentence that *must* have the probability r, given the aforementioned constraints. As an example that illustrates the method, the reader is invited to verify that the following sentence will always receive probability $\frac{1}{2}$ [the trick is to make extensive use of (PE9) and CCCP to deduce that the disjuncts in the antecedent, though disjoint, must have the same probability]:

$$[(\mathbf{C} \& (\neg \mathbf{C} \to (\mathbf{B} \& (\neg \mathbf{B} \to \mathbf{A})))) \vee (\mathbf{B} \& (\neg \mathbf{B} \to (\mathbf{C} \& (\neg \mathbf{C} \to \mathbf{A}))))]$$
$$\to (\mathbf{B} \& (\neg \mathbf{B} \to (\mathbf{C} \& (\neg \mathbf{C} \to \mathbf{A}))))$$

Stalnaker. Robert Stalnaker (1976) demonstrated in his "Letter to van Fraassen" that CCCP is inconsistent with the assumption that the logic of the "$\to$" is C2 (Stalnaker's preferred choice). That is, there are no non-trivial CCCP-functions for any algebra whose "$\to$" conforms to the axioms of C2. This is a strong result; however, there is a stronger result in the offing:

Weakened Transitivity Result. There are no nontrivial CCCP-functions for any algebra whose "$\to$" conforms to just (L1), (L2), and (L3). This result, of course, entails Stalnaker's result, because C2 includes all of (L1)–(L3). Here is a proof, whose leading idea – the choice of C – comes from Stalnaker's own:

Let $\langle W, F, P, \to \rangle$ be a model obeying CCCP; let (L1), (L2), and (L3) hold for the model. Suppose, for sake of *reductio*, that the model is non-trivial. Then there is a partition of W into three disjoint propositions, each of which has positive probability. Let three such be AB, $A\bar{B}$, and $\bar{A}$. Let $C = A \cup A \to B$. Then, referring now to the probabilistic entailment results, we have all of the following:

$$P(A \to C) = 1 \qquad \text{[by (PE1)]}$$
$$P(C \to AB) = P(C \to AB \cap C \to A) \qquad \text{[by (PE6)]}$$
$$P(C \to AB) = P(C \to AB \cap C \to B) \qquad \text{[by (PE6)]}$$

92

Therefore, $P(C \to AB) = P(C \to AB \cap C \to A \cap C \to B \cap A \to C)$. By (L3), it follows that $P(C \to AB) \leqslant P(C \to AB \cap A \to B)$. But $A \to B \subseteq C$; therefore $P(C \to AB) \leqslant P(C \to AB \cap C) = P(ABC)$, using here (PE9). Thus, by CCCP, $P(C \to AB) = P(AB|C) \leqslant P(ABC)$, whence $P(C) \geqslant 1$, whence $P(C) = 1$, whence $P(\bar{C}) = P(\bar{A} \cap A \to \bar{B}) = 0$ [relying here on (PE5)]. Therefore, using (PE9), $P(A\bar{B}) = P(A \cap A \to \bar{B}) = P(A \to \bar{B})$, whence, by CCCP and a rearrangement, $P(A\bar{B}) = P(A)P(A\bar{B})$, which is impossible. This completes the *reductio*.

This ends our overview of the triviality results. We turn now to a brief discussion of the conditions under which the Hypothesis can be maintained.

5.5. *Tenability results*

In the preceding section we focused on this question: What CCCP-functions are there for any given algebra? In this section it will be more convenient to ask, instead: What CCCP-conditionals are there for any given probability space? We begin with a fairly simple observation: There will be CCCP-conditionals for any probability space $\langle W, F, P \rangle$ in which the range of $P = [0, 1]$, at least when we do not demand any noteworthy logical structure of the conditional. Suppose we begin with such a probability space. Then for each $A, B \in F$ such that $P(A) > 0$ there will be a $C \in F$ such that $P(C) = P(B|A)$. For each such A and B, we choose such a C; we can then introduce an operation "$\to$" by the equation $A \to B = C$. The resulting model $\langle W, F, P, \to \rangle$ will automatically obey CCCP. Indeed, if for each A and B there are many choices of suitable C, then there will be correspondingly many choices of distinct CCCP-conditionals for the probability space. Let us call this fact the *Basic Tenability Result*.

Note that the requirement on the range of P in turn requires that W be infinite; thus, the most natural way to try to strengthen the Basic Tenability Result – by allowing W to be finite – is blocked by Hájek's Finitude Result. Furthermore, there is no guarantee that the CCCP-conditional chosen for a given probability space will behave at all like a *conditional* – that is, will obey any of (L1)–(L3). Indeed, we know by the Weakened Transitivity Result that it cannot obey all of (L1)–(L3), and we know by the No Atoms Result that it cannot even obey (L1) unless the original probability space is in fact *full*. Thus, the question naturally arises whether or not there *are* any models obeying CCCP and in which the "$\to$" really behaves like a conditional. Enter van Fraassen.

Van Fraassen's tenability results. In his 1976 paper, van Fraassen proves two results that demonstrate that the answer to the question just raised is "yes" – provided that one is willing to restrict the domain of propositions to which CCCP applies. That is, van Fraassen shows how to construct models $\langle W, F, P, \rightarrow \rangle$ in which "$\rightarrow$" does indeed behave like a conditional, and in which $P(A \rightarrow B) = P(B \mid A)$ for *some* pairs A, B such that $P(A) > 0$.

In order to make this claim more precise, we need to introduce a couple of new notions. First, we say that $\langle W^*, F^*, P^* \rangle$ *embeds* $\langle W, F, P \rangle$ just in case there is a sigma-field $F' \subseteq F^*$ and an isomorphism $g: F \rightarrow F'$ such that for all $A \in F$, $P^*(g(A)) = P(A)$. Intuitively, the probability space $\langle W, F, P \rangle$ can be "found within" any space that embeds it, and we can think of an element A of F as also belonging to F^* – where we mean by this that $g(A)$ belongs to F^*. In the same way we can speak of a subset of F as also being a subset of F^*, and for the sake of simplifying our exposition we shall so speak in what follows. Second, given any probability space $\langle W, F, P \rangle$, we will use the term *base set* to denote an arbitrary countable subset of F. (Heuristically, the propositions in a base set can be thought of as the denotations of a set of atomic sentences in some language.) Third, given any model $\langle W, F, P, \rightarrow \rangle$ and base set $S \subseteq F$, we call S^* the *extension* of S iff S^* is the closure of S under "$\rightarrow$" and the Boolean operations – *not including* countable union. Thus, the extension of a base set will be a field, but not a sigma-field. (Heuristically, the propositions in the extension of a base set can be thought of as the denotations of the sentences of some language equipped with a conditional.) Van Fraassen leads up to his first result by showing that any probability space can be embedded in a *full* space. He then proves the following:

Tenability Result for CE. Given any full probability space $\langle W, F, P \rangle$ and any base set S within it, there is an "$\rightarrow$" such that the model $\langle W, F, P, \rightarrow \rangle$ has the following features: (1) if S^* is the extension of S, then for all $A, B \in S^*$ such that $P(A) > 0$, $P(A \rightarrow B) = P(B \mid A)$; (2) the model obeys not only (L1) and (L2) but also all of the principles characteristic of the conditional logic CE.[19]

Tenability Result for C2. Given any probability space $\langle W_0, F_0, P_0 \rangle$ and any base set S within it, there is a model $\langle W, F, P, \rightarrow \rangle$ such that the following hold: (1) $\langle W, F, P \rangle$ embeds $\langle W_0, F_0, P_0 \rangle$; (2) for any $A, B, C \in S$, $P(A \rightarrow B) = P(B \mid A)$ and $P(A \rightarrow (B \rightarrow C)) = P(B \rightarrow C \mid A)$ if $P(A) > 0$, and $P((A \rightarrow B) \rightarrow C) = P(C \mid A \rightarrow B)$ if $P(A \rightarrow B) > 0$; (3) $\langle W, F, P, \rightarrow \rangle$ obeys all of the principles characteristic of the conditional logic C2.[20]

A number of comments are in order. To begin, van Fraassen understates each result: The proof of the first result in fact produces a model that obeys all of the principles characteristic of the conditional logic C2 save only weakened transitivity (call such a logic "C2−"); the proof of the second in fact guarantees that $P(A \rightarrow B) = P(B \mid A)$ provided just that $A \in S$

(i.e., B can be *any* element of F). Furthermore, we conjecture (although we have not proved) that the guiding idea behind the proof of the first result can be used to establish this stronger claim: For any full probability space $\langle W, F, P \rangle$, there is some "$\rightarrow$" such that the model $\langle W, F, P, \rightarrow \rangle$ obeys CCCP in *unrestricted* form, along with all of the principles of C2 −. Finally, it is useful to point out the various ways in which van Fraassen's positive results bump up against the negative results reviewed earlier. First, while it may be true (if our conjecture is correct) that any full space can be equipped with a CCCP-conditional that obeys C2 −, the Orthogonality Result guarantees that two distinct full spaces that differ only in the identity of the probability function cannot be equipped with the *same* CCCP-conditional unless the functions are orthogonal. Second, the Tenability Result for CE shows that so long as a model is full, it is possible for it to obey CCCP (perhaps in unrestricted form) *and* for its "$\rightarrow$" to have quite a bit of logical structure; the No Atoms Result, on the other hand, shows that even a minimum of logical structure *requires* that the model be full. Third, the Tenability Result for C2 shows that it is possible for a model to obey a very restricted form of CCCP while at the same time conforming to C2; the Weakened Transitivity Result, on the other hand, shows that these restrictions cannot be relaxed – for recall that the counterexample to CCCP used in the proof of that result had the form $[A \cup (A \rightarrow B)] \rightarrow AB$ (where A and B could be taken to belong to the base set S), involving only a slightly more complex antecedent than the allowed $(A \rightarrow B) \rightarrow AB$.

In somewhat more intuitive form, then, van Fraassen's results show that any probability space can be embedded within a model that obeys a restricted form of CCCP and whose "$\rightarrow$" does indeed behave like a conditional; greater "conditional-like" behavior is bought at the price of greater restrictions on the scope of CCCP.

This completes our exposition of the most significant technical material associated with the Hypothesis. In the remainder of this chapter we shall undertake a philosophical evaluation of the Hypothesis in light of these results.

6. WHY DISBELIEVE THE HYPOTHESIS?

Here we shall reconsider the various versions of the Hypothesis and whether or not they can withstand the onslaught of the many negative results canvassed in the preceding section; we shall consider, too, how much support they receive from van Fraassen's tenability results.

To begin, the universal version of the Hypothesis is clearly bankrupt, conclusively refuted many times over. We shall have no more to say about it.

95

Turning now to the belief function version, we can see at once that Lewis' results certainly scathe it. However, his results invite at least two responses from a defender of the Hypothesis: First, there might be a CCCP-conditional for *some* and perhaps *most* of the members of the class of belief functions – enough, at any rate, for CCCP to retain its interest. Second, the class of belief functions might not, in fact, be closed under either conditioning or Jeffrey conditioning. Let us consider these in turn.

In his 1976 paper, Lewis takes his first two results to be conclusive, giving the following argument (read "CCCP-conditional" for "probability conditional"):

> Even if there is a probability conditional for each probability function in a class, it does not follow that there is one probability conditional for the entire class. Different members of the class might require different interpretations of → to make the probabilities of conditionals and the conditional probabilities come out equal. But presumably our indicative conditional has a fixed interpretation, the same for speakers with different beliefs, and for one speaker before and after a change in his beliefs. Else how are disagreements about a conditional possible, or changes of mind? [p. 133]

Thus, Lewis believes that the interpretation of the conditional is independent of the beliefs of its utterer. But his argument to this conclusion – that a fixed interpretation is necessary for disagreement and changes of mind – is not fully convincing, and the easiest way to see why is to consider what other views might be refuted by such an argument. You and I are arguing about who is "over there"; the disagreement is a real one and is not due merely to confusion over the region to which we are referring. Does that mean that the meaning of "over there" as you use it, and as I use it, is independent of our beliefs? Of course not – nor does it mean that we happen at any rate to mean the same thing in *this* context. All that is required for substantive disagreement is that we mean *near enough* the same thing – that is, that the region that counts as "over there" for you sufficiently overlaps the region that counts as "over there" for me. Or again, if you and I are disagreeing about what is right or wrong to do in some situation, does it follow – given that our disagreement is substantive – that we mean exactly the same thing by "right" and "wrong"? Almost certainly not: If "right" means something like "sanctioned by my moral standards," then there is still plenty of room for real disagreement, provided that our standards overlap sufficiently.[21] Indeed, if Lewis' argument were sound, we would have ready to hand a quick and easy refutation of a number of kinds of ethical relativism. Too quick and easy, by our lights.[22]

Thus, the friend of the Hypothesis can dodge Lewis' argument that the "→" must be perfectly stable in interpretation. Then the spirit, if not the letter, of the belief function version might be preserved: There might be a very few arrows, very similar if not exactly alike in interpretation, such that each belief function would be a CCCP-function for at least one of them. Of course, such a maneuver leaves unfinished business: In particular, we need an explanation of what "very similar" means in this context.

Alternatively – turning now to the second response – one could insist that the set of belief functions is not closed under either conditioning or Jeffrey conditioning. One way to argue for this position would be via something stronger still: namely, that conditioning and Jeffrey conditioning always convert rational opinion to irrational. To some extent this captures Appiah's 1985 response to Lewis. There he argues that the probability function of a rational agent must be *regular* – it assigns probability zero only to the empty proposition – and observes that the result of conditioning is always an irregular probability function. Of course, because (non-degenerate) Jeffrey conditioning preserves regularity, Lewis' fourth result accommodates this point (as it was designed to), but perhaps the defender of the Hypothesis could augment Appiah's critique with an argument similarly directed against Jeffrey conditioning. If so, he would have a second way of evading Lewis' results.

Happily, we can now see that all such skirmishing is beside the point, thanks to the Orthogonality Result, for the belief function version of the Hypothesis is true only if any two distinct belief functions are orthogonal – and that is surely absurd, regardless of the status of conditioning and Jeffrey conditioning.[23] Nor will a more modest belief function version that requires only uniformity of the "→" over some such functions suffice, for this is still too much uniformity to avoid the Orthogonality Result. We conclude that the belief function version of the Hypothesis is untenable, as are modest weakenings thereof.

So, too, the universal tailoring version, as shown by Hájek's first result, for given any W and F, there will, of course, be functions definable on F whose ranges of values will be finite; as Hájek shows, for no such function P are there any CCCP-conditionals.

This leaves the belief function tailoring version: For each P that could represent a rational agent's system of beliefs, there is some CCCP-conditional. But we should note the constraints on this version, given various of the triviality results. To begin, let us suppose what might be controversial, that we can take the set of worlds W and the set of significant propositions F to be fixed, the same for any probability space representing a rational agent's beliefs.[24] Then we can take a model $\langle W, F, P, \rightarrow \rangle$ to be a combined representation of an agent's opinion together with her interpretation of the conditional. Now we can ask: Given that the belief function tailoring version of the Hypothesis is true, what are the models that represent rational agents like? The answer should make the defender of the Hypothesis uncomfortable: First, each belief function must have an infinite range (Hájek's first result). Second, if the conditional really behaves, even minimally, like a *conditional*, then each model must be full (the No Atoms Results). Third, there will be sentences of each agent's language – indeed, at least one for each rational probability value – whose probabilities will be fixed *purely by their syntactic form* (the Constructibility Result). Fourth, the conditional cannot have too much logical structure (Weakened Transitivity Result). Fifth, any two models whose probability functions differ without being orthogonal must interpret the conditional differently (the Orthogonality Result). These add up to quite a costly package; still, if the defender is willing to buy it, then he can have his Hypothesis (although possibly only in restricted form), thanks to van Fraassen's Tenability Result for CE. Let us now consider how steep the price tag is.[25]

Regarding the first and second points: We must emphasize that it is a *requirement* of the view we are considering that belief functions have infinite ranges, and most likely that they be full. One might indeed argue that, limitations on human discriminatory capacities and quantity of gray matter notwithstanding, infinitely ranged and even full probability functions are often appropriate for modeling rational human opinion. This suggestion is most plausible if we are considering the opinions of some highly idealized rational agent. But in either case, it is difficult to see how, in addition, rationality would *preclude* the appropriateness of finite or less-than-full functions (particularly when considering the highly idealized agent). The defender of the Hypothesis owes us an explanation of why this is so.

Regarding the third point: We have no trouble understanding how, by virtue of its syntax, a sentence might necessarily receive the probabilities 0 or 1. But how is it that syntax alone will guarantee that a sentence receives probability $\frac{17}{243}$? No "Principle of Indifference" has yet been advanced so bold, no Carnapian program yet carried out so far, as to deliver *this* result. Again, the defender owes us an explanation – either of this fact or of why the conditional lacks the minimal logical structure needed to generate it.

Regarding the fourth point: Stalnaker's own presentation of his result left the impression that it required the full strength of his logic C2, whence the result could be resisted simply by denying that C2 captures the structure of the relevant conditional. But the principles (L1)–(L3) are common to other conditional logics – Lewis' own being an example. The defender owes us an account of what the logic of the relevant conditional is, and of why – assuming that this logic yields the exceedingly plausible (L1) and (L2) – it does not yield a principle of weakened transitivity.

Regarding the fifth point: If, on the view we are considering, a rational agent changes her mind, then almost certainly her new belief function will fail to be orthogonal to her old one. Then her interpretation of the conditional must change – no matter how trivial her change of mind. Similarly, if your state of opinion and mine differ ever so slightly, then we interpret the conditional differently.[26] Indeed, van Fraassen, for one, seems willing to accept this claim. Rejecting Lewis' requirement of uniformity, he writes: "Would it not seem rather, that our probabilities are inextricably involved in the way we represent the possibilities, and nearness relations among them, to ourselves...if our ideas about the one change, will we not revise our modeling of the other?" (1976, p. 274). It appears that on van Fraassen's view the conditional is so intimately tied to opinion that *any* change in opinion entails corresponding changes in which propositions are picked out by conditional utterances. Note just *how* radical this is: You previously assigned probability $\frac{1}{2}$ to the coin landing heads; but now that you have seen that it landed tails, you assign that proposition probability zero – with consequent revisions in how the conditional functions for you. Indeed, this feature of the belief function version of the Hypothesis imposes the costliest burdens on its defender. Consider just some of what needs to be explained:[27]

1. Why must apparently trivial changes of opinion change one's interpretation of the conditional?
2. *How* do changes of opinion change the interpretation of the conditional? That is, is this change rule-governed, and if so, what are the rules?
3. You and I can disagree about a conditional sentence, even though we must mean different things by it. Is this possible because we mean near enough the same thing? What does "near enough" *mean*, in this context?
4. The view we are considering can be characterized as claiming that the conditional acts like an indexical, where virtually any difference in the opinions of the speaker counts as a difference in index. For *other* indexicals ("here," "you," etc.) there are easily specifiable rules that take us from a context of usage, or index, to a meaning. What are the corresponding rules for the conditional?
5. Given the sensitivity of what is said in a conditional utterance to the degrees of belief of the sayer, why are the resources of ordinary English to indicate such degrees of belief so impoverished? Compare the case of "here," where we have rich resources for guaranteeing that our interlocutors know exactly where we are.
6. If the content of what one says in uttering a conditional is fixed in large part by one's state of opinion, then communication involving conditionals will either be almost impossible (if the interlocutor's knowledge of one's state of opinion is too incomplete) or be pointless (if it isn't).

We seriously doubt that the challenges listed here can be met; degrees of belief just do not seem to be relevant to what is said by utterances of conditionals – certainly not to the extent necessary in order to maintain the interest of the belief function tailoring version of the Hypothesis. We conclude that even this version – though it is the only one of those we have identified that has any chance of standing up to the triviality results – is not, in the final analysis, tenable.

7. CAN THE HYPOTHESIS BE SAVED?

The diehard supporter of the Hypothesis might now object that the four versions we have elucidated in this chapter far from exhaust the ways of giving precise content to the idea that conditional probability is the probability of some conditional. We certainly agree – provided that it is recognized that any available alternative must be substantially weaker than the four versions discussed, and thus correspondingly less interesting. Nevertheless, we shall in this section offer some suggestions of various ways in which the Hypothesis might be resuscitated, along with brief comments on problems facing each such maneuver. We consider four major options:

7.1. Deny that the "probability" of the conditional is really probability

The Hypothesis can be saved, in some form, if we replace the *probability* of the conditional $A \rightarrow B$ with some other quantity associated with $A \rightarrow B$. Reading Adams as many authors have, we might take this quantity to be the *assertability* of the conditional. Or we might consider the proposal of Stalnaker and Jeffrey (Chapter 4, this volume): Developing an idea due to de Finetti (1936), they treat sentences as random variables – functions from worlds to real numbers – and replace "probability" by "expectation value." In a very interesting result that synthesizes quite a lot of literature, they show how to develop the random variables approach so as to guarantee, for a wide range of conditionals $A \rightarrow B$, that the expectation of $A \rightarrow B$ equals the conditional expectation of B on A.

We have little to say about these approaches and shall confine ourselves to considerations of a very general sort: Either the quantity that replaces probability has some relevance to the actual use we make of conditionals in our thought and speech, or it hasn't. In the case of Stalnaker and Jeffrey, we simply ask if their expectation of sentences as random variables does have such relevance, and in what it consists. We leave this as an open question – and, to be fair, it is not the purpose of Stalnaker and Jeffrey's chapter to address it. On the other hand, Adams' Thesis as it is usually understood (which, we remind the reader, is not how Adams understands it!) makes use of a quantity that clearly is relevant, because we can understand assertability to be a measure of the degree of "appropriateness" or "warrantedness" of an utterance in a given context. But there are problems with his hypothesis that the assertability of a conditional goes by the conditional probability, if assertability is understood in this familiar way. If "goes by" means "exactly equals," then by Hájek's first result assertability turns out to be incredibly fine-grained, because the range of assertability values for a given agent will either be infinite or at any rate be greater than the range of values of her (unconditional) opinion. Such fine-grainedness is surely surprising, and it seems to stand in need of independent justification. If "goes by" means something much weaker – for example, that the assertability of a conditional is high, middling, or low according as the conditional probability is high, middling, or low – then the Hypothesis becomes more plausible, but less interesting; in particular, it seems unlikely that it could play a very useful role in the development of probability logic, or in other philosophical projects.[28]

Let us move on, then, to three quite different and related strategies for reviving the Hypothesis.

7.2. Restrict the domain of propositions

Perhaps it was unreasonable to hope that conditional probability would equal the probability of a conditional for *all* possible antecedents and consequents. That lets in propositions that may have no simple expression in a given language – for example, multiply iterated conditionals, extravagant Boolean combinations of conditionals, and so on – and such sentences are never uttered in natural language. And it is reasonable to think that all the linguistic data that we can garner are unable to distinguish between the full-blown Hypothesis and the Hypothesis with some suitable restriction on the domain – thus, the latter could suffice to explain why reports of conditional probabilities and probabilities of conditionals seem to say the same thing. Of course, we shall not get many of the other goodies that the full-blown Hypothesis was supposed to deliver (not in unrestricted form, at least): interesting constraints on the semantics of the conditional, insights into the theory of confirmation, help in constructing a "probabilistic soundness" criterion for arguments, solutions to interesting problems in probability kinematics, and so forth. But perhaps we should take what we can get, and accept

Restricted CCCP: $\qquad P(A \to B) = P(B \mid A) \quad$ for all $A, B \in \mathbf{S}$

where $\mathbf{S}$ is some proper subset of F.

101

Restricted CCCP yields different versions of what we might call "the Restricted Hypothesis." For example, the belief function version of the Restricted Hypothesis says that there is some "→" such that for all probability functions that could represent rational belief, Restricted CCCP holds. There are payoffs: If we accept Restricted CCCP with appropriate choice of S, then – thanks to van Fraassen's Tenability Result for C2 – we can also let the logic of the "→" be as strong as C2. But there are also serious problems. Because these problems are very similar to ones facing the next two fallback positions, we shall postpone their discussion until those positions have been laid out.

7.3. Retreat to an approximate version of CCCP

Instead of demanding the exact equality of conditional probability with the probability of a conditional, we might demand only that they *approximately* equal each other:

Approximate CCCP: $$P(A \rightarrow B) \approx P(B \mid A)$$

for all A, B in the domain of P such that $P(A) > 0$

We can then amend the original versions of the Hypothesis, replacing the exact CCCP with Approximate CCCP. Call these amendments versions of "the Approximate Hypothesis."

We have not yet said how "approximately" is to be understood – how the "$\approx$" is to be read. Two readings suggest themselves:

Reading 1: $P(A \rightarrow B) \approx P(B \mid A)$ iff $P(A \rightarrow B) = kP(B \mid A)$, where k is sufficiently close to 1.

Reading 2: $P(A \rightarrow B) \approx P(B \mid A)$ iff $P(B \mid A) - \varepsilon < P(A \rightarrow B) < P(B \mid A) + \varepsilon$, where ε is sufficiently close to zero.[29]

The retreat to the Approximate Hypothesis could be principled. Echoing reasons given earlier, its proponent might insist that all the available linguistic data are unable to distinguish between the (exact) Hypothesis and the Approximate Hypothesis – at least if the approximation is sufficiently good. Still, as we hope to make clear at the end of the next section, the Approximate Hypothesis will face some serious problems.

7.4. Cast the Hypothesis in terms of vague probability

Suppose we drop the presupposition, shared by all the belief-related versions of the Hypothesis that we have met so far, that a system of belief can be represented by a single function. Instead, let us insist that opinion should in general be represented as *vague*. Doing so makes room for yet another weakened version of the Hypothesis, but before getting to it we must take a brief detour through the literature on vague opinion.

For many purposes, it suffices to introduce interval-valued probability functions to represent vague opinion, but we shall follow van Fraassen's (1990) representation, which in turn follows proposals by Levi (1980) and Jeffrey (1983). Consider the set of all probability functions that are consistent with your determinate judgments – that correspond to all the arbitrary ways of "precisifying" your opinion, compatible with the state of mind that such judgments express. Call this set your "Representor." Suppose, for example, that the probability that you assign to Collingwood winning the Grand Final is not a sharp value; rather, it lies in some interval, say [0.7, 0.9].[30] Then your representor contains all probability functions that assign a sharp value in the interval [0.7, 0.9] to Collingwood winning and that are compatible with your other judgments. Now, given any set of functions $\{P\}$ that could serve as the representor for a rational agent's opinion, we can let "$\{P(A)\}$" stand for the agent's opinion about an arbitrary proposition A. We can now state a new restriction on CCCP:[31]

Vague CCCP. For all A, B in the domain of a representor's functions such that $0 \notin \{P(A)\}$, $\{P(A \to B)\} = \{P(B|A)\}$.

As before, Vague CCCP gives rise to different versions of what we might call "the Vague Hypothesis," with quantification over probability functions replaced by quantification over representors. And as before, one could argue that linguistic data do not distinguish between the Vague Hypothesis and the Hypothesis – given what must be argued, that ordinary human opinion is almost always at least a little bit vague. But, as with the two weakenings introduced earlier, there are problems.

These problems center around the Strengthened Lewis Result and Hájek's Perturbation Result, both of which can be recast so as to spell trouble for the three weakenings of the Hypothesis just discussed. We take them in turn:

Suppose that Restricted CCCP holds for two nontrivial models $\langle W, F, P, \to \rangle$ and $\langle W, F, P_C, \to \rangle$ that employ the same arrow, and such that P_C is derived from P via conditioning on a proposition C. Then the Strengthened Lewis Result guarantees that the restriction must exclude all propositions of the form $E \to \bar{C}$, where $E = \bar{C} \cup D$ for some $D \subset C$ such that $0 < P(D) < P(C)$. Again, if the Restricted Hypothesis holds for a nontrivial model $\langle W, F, P, \to \rangle$, then, for any choice of A and B that is not ruled out by the restriction, the Perturbation Result guarantees that there will be a wide range of models – those whose probability functions are *perturbations* of P relative to "$\to$," A, and B – for which Restricted CCCP fails.[32] Any restriction that works only by ruling out overly complicated compounds of conditionals will run afoul of these results.

Next, suppose that Approximate CCCP holds for two nontrivial models $\langle W, F, P, \rightarrow \rangle$ and $\langle W, F, P_C, \rightarrow \rangle$, as before. Choose E as before. Then, recalling that there are two readings of "$\approx$," straightforward modifications of the Strengthened Lewis Result[33] reveal the following: on Reading 1, $P(E) \geqslant k$; on Reading 2, $P(E) \geqslant P(\bar{C})/[2\varepsilon + P(\bar{C})]$. Given that k is sufficiently close to 1 and ε is sufficiently close to 0, $P(E)$ will be close to 1, whence it follows that for any $D \subset C$, $P(D) \approx P(C)$. But this is impossible [unless $P(C) \approx 0$]. Thus – unless the standards of approximation are so weak as to make the claim trivial – the Approximate Hypothesis falls prey to the Strengthened Lewis Result. It likewise suffers at the hands of the Perturbation Result, for Hájek's method for constructing counter-examples to CCCP can in fact be used to find drastic counterexamples: Specifically, for many nontrivial models $\langle W, F, P, \rightarrow \rangle$ and A, B such that $P(A \rightarrow B) = P(B|A)$, there will be a range of models $\langle W, F, P^*, \rightarrow \rangle$ such that P^* is a perturbation of P, and the values $P^*(A \rightarrow B)$ and $P^*(B|A)$ differ by as wide a margin as is desired.

Finally, suppose that Vague CCCP holds for some representor. We can analyze this as follows: Let the representor $\mathbf{R}$ be a class of probability functions defined on a particular algebra $\langle W, F, \rightarrow \rangle$. Then Vague CCCP states that $\{P(A \rightarrow B)|P \in \mathbf{R}\} = \{P(B|A)|P \in \mathbf{R}\}$ as long as $0 \notin \{P(A)|P \in \mathbf{R}\}$. In other words (given the condition on A), for each x in $[0, 1]$ there is a $P \in \mathbf{R}$ such that $x = P(A \rightarrow B)$ iff there is a $P' \in \mathbf{R}$ such that $x = P'(B|A)$. Now, for any proposition A, we can define the *spread* of A to be the least upper bound of the set $\{x | \text{for some } P, P' \in \mathbf{R}, x = |P(A) - P'(A)|\}$. Thus, if $\mathbf{R}$ represents a state of opinion that is perfectly precise with respect to A, then the spread of A is zero. Similarly, let the spread of $\mathbf{R}$ itself be the least upper bound of the set $\{x | \text{for some } A, x \text{ is the spread of } A\}$. We can thus think of the spread of $\mathbf{R}$ as a very crude measure of the level of vagueness in the state of opinion that it represents.

Now, letting ε be the spread of $\mathbf{R}$, it follows at once that each $P \in \mathbf{R}$ obeys Approximate CCCP for that ε (adopting Reading 2). Then, if ε is small enough, the Strengthened Lewis Result and Perturbation Result apply, in two ways: First, various shifts of opinion that take the subject from one representor to another will be ruled out (provided that Vague CCCP is supposed to hold before and after the shift, and that the shift does not increase the spread of the representor). Second, the composition of $\mathbf{R}$ itself will be severely limited (e.g., because various perturbations of members of $\mathbf{R}$ cannot be members of $\mathbf{R}$). These consequences spell trouble, and though a requirement that any acceptable representor have a spread greater than some fixed minimum might protect the Vague

Hypothesis from this trouble, we do not see how such a *general* requirement of vagueness – as opposed to a requirement that would mandate vagueness only in particular rare cases[34] – could be justified. Indeed, because the triviality results acquire more bite as the spread decreases, the proponent of this version of the Hypothesis is forced to conclude that each sharpening of opinion is a step in the direction of irrationality, which is surely ridiculous.

Our discussion of the possible alternatives to the full-blooded Hypothesis will stop here. We must emphasize that, far from providing a complete survey of all such strategies for weakening it, we have not even provided verdicts on all versions of those strategies we have included. For example, nothing that we have said counts against the universal tailoring version of the Approximate Hypothesis – we have recast two of the limitation results here, but not the no-go results. Still, we hope that by now it is overwhelmingly plausible that any defender of the Hypothesis – weakened or otherwise – has his work cut out for him. Only more investigation will tell whether or not there is some tenable and useful weakened form of the Hypothesis. In the final section, we consider some reasons for thinking that such investigation would be worthwhile.

8. CONCLUSIONS AND OPEN QUESTIONS

The most ambitious and most interesting forms of the Hypothesis are – with the possible exception of the belief function tailoring version – failures. But these failures leave us with an open question: What exactly is the relationship between conditional probabilities and probabilities of conditionals? Here we shall very briefly consider various responses to this question. It will become clear that we should not be satisfied simply because we now know a great deal about what this relationship is *not*.

On the one hand, one might respond that there is a fairly close relationship: In very many cases the two quantities coincide, or at least nearly so, for some particular choice of conditional. Such a position explains admirably the observation that the Hypothesis "sounds right," and it conspires happily with a view that such phrases as "the probability that if A, then B" are systematically ambiguous (in a way that ordinarily makes little or no difference) between two readings – $P(A \to B)$ and $P(B|A)$ – that we now know must be distinct.[35] But to endorse such a position just *is* to endorse some weakened version of the Hypothesis, most likely a version of the Restricted Hypothesis. As we saw in the preceding section, it remains to be seen whether or not there is a version of the Hypothesis weak enough to be tenable while strong enough to do justice to this response.

On the other hand, one could claim that there is no relationship of interest, that conditional probabilities and probabilities of conditionals vary quite independently of each other, that they coincide seldom and only by accident, and that any appearance to the contrary can be explained away. One way of doing some of the explaining away is found in the position of Lewis (1976) and Jackson (1987): The indicative conditional "If A, then B" can be correctly analyzed as having the truth conditions of the material conditional "$A \supset B$"; nevertheless, considerations from pragmatics show that one will typically assert it only if one's subjective conditional probability $P(B|A)$ is sufficiently high. The conditional probability then serves as a measure of the assertability of the indicative conditional, but not of its probability. But while this maneuver successfully explains away one of the initial motivations for the Hypothesis (Adams' Thesis), it does not explain enough (not by itself, at least). For example, why does the honest response to a question of the form "How likely is it that if A, then B?" seem to be to report one's estimate of the conditional probability of B, given A, as opposed to the probability of the material conditional?[36] (Indeed, the kind of reliance on pragmatics that features so prominently in the positions of Jackson and Lewis must be carefully watched, lest it make the chosen truth conditions irrelevant. After all, one would not want to hold, for example, the position that the truth conditions for the indicative were the same as those for the contradiction – explaining away all appearances to the contrary by pragmatics.)

Furthermore, there are other conditionals to consider besides the indicative. What of future-directed counterfactuals? Suppose you ask, "Were Perot to run again in 1996, would he win?" We respond: "Dunno. It seems just as likely as not." Taking the "$\rightarrow$" to be the counterfactual, have we just reported that, for us, $P(\text{Perot runs} \rightarrow \text{Perot wins}) = \frac{1}{2}$, or rather that $P(\text{Perot wins} | \text{Perot runs}) = \frac{1}{2}$? Perhaps *both*; indeed, it seems that in very many cases, the probability of such a future-directed counterfactual should equal the conditional probability. No discussion that confines itself to the indicative can explain why.

Surprisingly, this claim of significant overlap gets support from causal decision theory – at least, on the formulation due to Gibbard and Harper (1978). This formulation can be arrived at by first formulating *non*-causal decision theory in a certain way, and then, at a key point, replacing conditional probability by the probability of the corresponding counterfactual conditional. If the Gibbard/Harper formulation is adequate, then to the extent that causal and non-causal decision theories deliver the same results – which decision theorists take to be most of the time – these quantities should be equal throughout a substantial range of cases.

Finally, it is worth remembering other reasons, yet to be addressed, for holding the Hypothesis, most notably the last of those discussed in

106

Section 3 ("Independence of the conditional from its antecedent"). If there is no significant overlap between conditional probabilities and probabilities of conditionals, then, given any conditional that obeys centering (e.g., the counterfactual, according to Lewis, Stalnaker, and others), it should be obvious that conditionals of that sort are typically *not* probabilistically independent of their antecedents. But this is not at all obvious.

Of course, we have not considered in depth the responses that could be given to the question posed at the beginning of this section, nor is it possible for us to do so in this chapter. But we commend more detailed consideration of this question as a worthy philosophical task, for by now it should be clear that the failure of the Hypothesis in its original forms leaves us with plenty of work to be done. In a way this is a happy ending for the Hypothesis, for though its demise may have stymied certain research programs, that should only give impetus to others.[37]

NOTES

1. Stalnaker extends the classical probability calculus to one in which conditional probabilities are primitive, so that conditional probabilities are defined even when the condition has probability zero; however, this falls outside the scope of our CCCP.

   Strictly speaking, we should distinguish propositions from the sentences that express them, and we should make clear whether we are thinking of probability as attaching to the former or to the latter. We shall not fuss about this distinction when nothing turns on it, and we take this to be common practice.
2. Page references are to Harper et al. (1981) for all articles that are cited as appearing there.
3. Jeffrey conditioning was first introduced by Jeffrey (1965) under the name "probability kinematics."
4. This is the motivation for the Hypothesis discussed by Lewis (1976, 1986).
5. Usual, but not unproblematic: See the next section for worries about centering. Note that "centering" is a slight misnomer, since this name usually refers to a property of the nearness relation used to give the truth conditions for the conditional (each world is the nearest world to itself). However, since this property yields the logical principle used in the text, no harm is done in calling this principle by the same name.
6. Note that if $P(A) = 0$, the equation holds trivially.
7. Stalnaker believes that this "→" is in fact the Stalnaker conditional. There is really no issue for him of *which* conditional is involved here, since it is part of his position that his is the only one – although, to be sure, he does regard the indicative/subjunctive distinction of grammar as playing a role in determining a certain contextually determined parameter, the selection function, which in turn partly determines the conditional's truth conditions.
8. We find these assumptions highly plausible for any connective that deserves to be called a "conditional." But they are not entirely uncontroversial: For example, McGee (1985) disputes the unrestricted use of *modus ponens*.

107

9. The deduction is given in Section 5.3. Note that the logical principle of centering is *not* required.
10. We are indebted to Bas van Fraassen here.
11. It follows that P has at most four *un*conditional values. However, the reverse implication does not hold: Consider a model in which W contains just three worlds, each given weight $\frac{1}{3}$.
12. Here's how: Note first that triviality guarantees that for all $X, Y \in F$, the value of $P(X \cap Y)$ is $0, 1, P(X)$, or $P(Y)$. Then for all $X, Y \in F$, let $X \to Y = \varnothing$ if $P(X \cap Y) = 0$; otherwise, let $X \to Y = W$ if $P(X \cap Y) = 1$ or $P(X)$; otherwise, let $X \to Y = Y$.
13. For the sake of precision, we use set-theoretical notation throughout this section. We also frequently suppress the set intersection sign.
14. The reverse implication – deriving CCCP from the independence condition – needs more than (L1) and (L2). This can be seen by supposing that, for all A and B, $A \to B = \varnothing$. Then (L1), (L2), and the independence condition all hold, but not CCCP.
15. The restriction on x guarantees that we do not simply have a case of conditioning. Note that this formulation, which Lewis uses in the proof of his fourth triviality result, is equivalent to $P_x(B) = \alpha P(B|C) + (1 - \alpha)P(B|\neg C)$, with $\alpha = P(C) + x$. This may be a more familiar schema for Jeffrey conditioning over a two-celled partition; α represents the new probability assigned to C, after a learning experience that makes it more probable by an amount x than it was previously.
16. Note also that there are various equivalent ways of stating the results: For example, we might state the second result by saying that for any class of models closed under conditioning, and over which the conditional is interpreted uniformly, CCCP holds throughout the class only if it contains just trivial models.
17. A proof of the equivalence between "full" and "atomless" is given in the Appendix to Hall's Chapter 8.
18. As Hall points out in his discussion of this result, these atomic sentences can be replaced by nonconditional sentences whose syntax guarantees that they must be interpreted as disjoint. The sentences $(A \& B)$, $(A \& \neg B)$, and $\neg A$ will do, for example.
19. We shall not rehearse these principles here; readers are referred to van Fraassen's paper for details.
20. Again, we shall not rehearse these principles; but recall that C2 includes the weakened transitivity condition (L3).
21. We owe this point to Jamie Dreier.
22. Lewis' assumption that the "$\to$" has uniform interpretation, irrespective of the probability function in whose scope it appears, is dubbed "metaphysical realism" by Stalnaker (1976) – an allusion to the Lewis metaphysical framework, no doubt, although the assumption could be found plausible without it. Van Fraassen (1976) surmises that Lewis' framework – his realism about possible worlds, and about the similarity relations thereon – affords him an argument for the uniformity of "$\to$". The idea is that there is an objective fact about what the worlds are, and an objective fact about what the similarity relations among them are, and that together these determine the truth values of all propositions. The rational agent's uncertainty about which world she inhabits is represented by her probability assignment to the worlds; but the question

which worlds constitute a given proposition – even a proposition expressed by a conditional – is insensitive to this assignment. Note, however, that van Fraassen's formulation of the argument seems to assume that Lewis must regard "→" as depending on the constituents of his metaphysical framework: yet Lewis' interest in the Hypothesis really concerns the indicative conditional, which he believes to be simply the material conditional. More importantly, van Fraassen is unfair to Lewis when he writes that his uniformity assumption is "justified only by metaphysics" (p. 275) – after all, Lewis' argument quoted earlier makes no appeal to his metaphysics and could be found persuasive by someone who does not subscribe to that metaphysics.

23. Note, too, that regularity is violated in a big way – unless there is only one belief function!

24. This should at least be the case for the set of spaces that record the epistemic history of a single agent. We shall not stop to argue for the stronger claim here.

25. Our discussion here will be brief, substituting sketches for fully fleshed-out arguments. For those, the reader is referred to Hájek (1993).

26. This is perhaps too quick: One might argue that if the only difference can be fully described by noting that you are certain that p, while I am certain that $\neg p$, then that difference is still slight, even though it yields orthogonal belief functions. Needless to say, granting this point does almost nothing to undermine the arguments that follow.

27. The points that follow are given extensive treatment by Hájek (1993).

28. See Hájek (1993) for more discussion of the points raised in this paragraph; see Collins (1991) for considerations that tell strongly against fine-grainedness for assertability.

29. Note that Reading 1 implies Reading 2, for if $P(A \to B) = kP(B|A)$ and for some small ε, $1 - \varepsilon < k < 1 + \varepsilon$, then $P(B|A)(1 - \varepsilon) < P(A \to B) < P(B|A)(1 + \varepsilon)$, whence $P(B|A) - \varepsilon < P(A \to B) < P(B|A) + \varepsilon$.

30. Probably the endpoints of this interval should be vague as well, but we shall overlook that complication.

31. Michael Thau and Bas van Fraassen, both playing the role of devil's advocate, independently suggested this alternative to one of us. Note that there are other, somewhat less plausible ways to incorporate vagueness into a weakened version of the Hypothesis. Space considerations preclude their discussion here; however, see Hájek (1993) for extensive discussion.

32. Both of these claims would have to be weakened if the content of the restriction were allowed to vary from model to model. We shall not pursue that possibility here.

33. Set forth in Hájek (1993).

34. For example, if you have full knowledge of the objective chances regarding the outcome of some indeterministic process, and those chances are *themselves* quite vague, then so, perhaps, should be your opinion about the outcome. Also, Bas van Fraassen has brought to our attention situations arising in quantum mechanics where, after conditioning on a certain proposition A (which fully describes both the measurement outcome and apparatus setting for one half of a Bell's Inequalities setup), one's opinion about a different proposition B (which describes the outcome of the other half of the setup) has gone from perfectly sharp to completely vague. However, these turn out to be situations in which one's opinion about some further proposition C (which

describes the apparatus setting for the other half) is completely vague both before and after the shift.

35. Hájek (1992, 1993) explores this possibility at much greater length.

36. The idea that such a request might reasonably be answered with an estimate of $P(A \supset B)$ seems to us to lead to amusing absurdities. For example, suppose we are haggling about where to go for dinner; we want to go to Le Snob, while you, thinking it overpriced, want to go to Burger King. We don't have reservations for either place, but find this more worrying in the case of Le Snob. By way of coming to decision, we ask you, "How likely is it that if we go to Le Snob, we won't be able to get a table?" You, having no idea how crowded Le Snob is tonight, nevertheless respond, "Very likely," and so we don't go. Have you been deceitful? Of course, but not according to the rule in question: for given that you knew you were going to say "Very likely," your subjective probability P(we don't go) was very high, hence so too your subjective probability P(we go $\supset$ we don't get a table). A nifty trick!

37. We wish to thank the following people for extensive and helpful discussions: John Barker, Alex Byrne, Fiona Cowie, Bas van Fraassen, Dick Jeffrey, Mark Kalderon, David Lewis, Jennifer Saul, Brian Skyrms, Michael Thau, and Lyle Zynda.

REFERENCES

Adams, Ernest (1965) "The Logic of Conditionals." *Inquiry* 8:166–97.
Adams, Ernest (1975) *The Logic of Conditionals*. D. Reidel, Dordrecht.
Appiah, Anthony (1985) *Assertion and Conditionals*. Cambridge University Press.
Collins, John (1991) "Belief Revision." Ph.D. dissertation, Princeton University.
de Finetti, Bruno (1936) "La Logique de la Probabilité." *Induction et Probabilité, Actualités Scientifiques et Industrielles* 391:31–9.
de Finetti, Bruno (1972) *Probability, Induction and Statistics*. Wiley, New York.
Ellis, Brian (1969) "An Epistemological Concept of Truth." Pp. 52–72 in *Contemporary Philosophy in Australia*, ed. Robert Brown & C. D. Rollins. Allen & Unwin, London.
Gibbard, A., & Harper, W. L. (1978) "Counterfactuals and Two Kinds of Expected Utility." Pp. 125–62 in *Foundations and Applications of Decision Theory*, ed. C. A. Hooker, J. J. Leach, & E. F. McClennen. D. Reidel, Dordrecht.
Hájek, Alan (1989) "Probabilities of Conditionals – Revisited." *Journal of Philosophical Logic* 18:423–8.
Hájek, Alan (1992) "A Dogma of Conditional Probability." Unpublished manuscript.
Hájek, Alan (1993) "The Conditional Construal of Conditional Probability." Ph.D. dissertation, Princeton University.
Harper, W. L., Stalnaker, R. & Pearce, G. (eds.) (1981) *Ifs*. D. Reidel, Dordrecht.
Jackson, Frank (1987) *Conditionals*. Blackwell, Oxford.
Jeffrey, Richard (1964) "If" (abstract). *Journal of Philosophy* 61:702–3.
Jeffrey, Richard (1965) *The Logic of Decision*. University of Chicago Press.
Jeffrey, Richard (1983) "Bayesianism with a Human Face." Pp. 77–107 in *Probability and the Art of Judgment*. Cambridge University Press.
Jeffrey, Richard (1992) *Probability and the Art of Judgement*. Cambridge University Press.
Joyce, James (1991) "The Axiomatic Foundations of Bayesian Decision Theory." Ph.D. dissertation, University of Michigan.
Levi, Isaac (1980) *The Enterprise of Knowledge*. MIT Press, Cambridge, Mass.
Lewis, David (1971) "Completeness and Decidability of Three Logics of Counterfactual Conditionals.." *Theoria* 37:74–85.

Lewis, David (1973) *Counterfactuals.* Blackwell, Oxford.

Lewis, David (1976) "Probabilities of Conditionals and Conditional Probabilities." *Philosophical Review* 85:297–315; reprinted in Harper et al. (eds.) (1981) *Ifs.* D. Reidel, Dordrecht.

Lewis, David (1986) "Probabilities of Conditionals and Conditional Probabilities II." *Philosophical Review* 95:581–9.

McGee, Vann (1985) "A Counterexample to Modus Ponens." *Journal of Philosophy* 82: 462–70.

Quine, Willard Van Orman (1950) *Methods of Logic.* Holt, New York.

Ramsey, Frank Plumpton (1965) *The Foundations of Mathematics (and Other Logical Essays).* Routledge & Kegan Paul, London.

Stalnaker, Robert (1968) "A Theory of Conditionals." *Studies in Logical Theory. American Philosophical Quarterly* monograph series, no. 2. Blackwell, Oxford.

Stalnaker, Robert (1970) "Probability and Conditionals." *Philosophy of Science* 37:64–80, reprinted in Harper et al. (eds.) (1981) *Ifs.* D. Reidel, Dordrecht.

Stalnaker, Robert (1976) "Letter to van Fraassen." Pp. 302–6 in *Foundations of Probability Theory, Statistical Inference, and Statistical Theories of Science,* vol. 1, ed. W. L. Harper & C. A. Hooker. D. Reidel, Dordrecht.

van Fraassen, Bas (1976) "Probabilities of Conditionals." Pp. 261–301 in *Foundations of Probability Theory, Statistical Inference, and Statistical Theories of Science,* vol. 1, ed. W. L. Harper & C. A. Hooker. D. Reidel, Dordrecht.

van Fraassen, Bas (1989) *Laws and Symmetry.* Clarendon Press, Oxford.

van Fraassen, Bas (1990) "Figures in a Probability Landscape." Pp. 345–56 in *Truth or Consequences,* ed. J. M. Dunn & A. Gupta. Kluwer, Dordrecht.

7

Triviality on the cheap?

ALAN HÁJEK

1. PRELIMINARIES

The hypothesis that conditional probability equals the probability of some conditional (hereafter, "the Hypothesis") is not dead yet. It can thread its way through the loopholes left open by the triviality results of Lewis, Stalnaker, and others. Furthermore, a positive result due to van Fraassen (1976) ensures its immortality in a certain sense, which we shall discuss. Thus we know that in that sense, the search for knockdown results against it is doomed to fail. Nevertheless, the Hypothesis was originally proposed to play a role in providing a semantics for the conditional – it was thought by Stalnaker (1970) to be a criterion of adequacy for such a semantics, and a variant of it was employed by Adams (1975) in his account of the indicative conditional. In Chapter 6 of this volume, Ned Hall and I argue that the version of the Hypothesis that van Fraassen has immortalized is not fit for this task. So interest remains in the tenability of some form of the Hypothesis that can plausibly be construed as a semantic rule for the conditional, and as such, it is still fair game. With this in mind, I shall offer here another set of results against it. They have their loopholes too – but I think that they are not so easily threaded.

It will help to focus our discussion if we quickly review some terminology and distinctions that were introduced by Hájek and Hall in Chapter 6. Let W be a set, F a sigma-algebra of subsets of W, and P a probability function on F. Think of W as a set of worlds, and F as the set of significant propositions that are assigned probabilities by P. Let "$\rightarrow$" be some function from pairs of propositions to propositions. Of course, our interest is really in an "$\rightarrow$" that has some claim to forming conditionals, but this will not be essential. Call $\langle W, F, P, \rightarrow \rangle$ a *model*.

Now we can introduce the equation that encapsulates the conditional construal of conditional probability:

CCCP: $\qquad P(A \rightarrow B) = P(B|A) \quad$ for all A, B in F, with $P(A) > 0$

CCCP asserts that a certain relationship holds between P and "→." When it does, we may say either that "→" is a *CCCP-conditional for P* or that P is a *CCCP-function for* "→." Call a function that is not a CCCP-function for "→" a *non-CCCP-function for* "→." (I shall sometimes drop the reference to the "→," when it is clear which one I mean.) Call P a *belief function* if P could represent a rational agent's system of beliefs; let us not fuss about exactly what that might require.

We generate versions of the Hypothesis by quantifying in various ways over the probability functions and the "→" that appear in CCCP. Several versions are distinguished by Hájek and Hall (Chapter 6), but only two of them will be needed here. Given a prior specification of W and F, they are as follows:

Universal version: There is some "→" such that CCCP holds for all models $\langle W, F, P, \to \rangle$.

Belief function version: There is some "→" such that CCCP holds for all models $\langle W, F, P, \to \rangle$ where P is a belief function.

For reasons that are given by Hájek and Hall in Chapter 6, neither of these versions of the Hypothesis is tenable (nor are various other versions that we introduce). Nevertheless, there remain strategies for resuscitating the Hypothesis. In this chapter I want to discuss – and thwart – one strategy that strikes me as particularly salient.

It was perhaps a vain hope that conditional probability should equal the probability of a conditional for *all* possible antecedents and consequents. That lets in propositions that may have no simple expression in a given language – for example, multiply iterated conditionals, extravagant Boolean combinations of conditionals, and so on – and we never utter such sentences. Indeed, all the linguistic data that we can garner are unable to distinguish between the full-blown Hypothesis and the Hypothesis with some suitable restriction on the domain. But for that very reason, the latter could suffice to serve our purposes in illuminating the semantics of the conditional, and it is reasonable to hope that it *is* tenable.

So we might try shrinking the domain of propositions that appears in the statement of CCCP. After all, domain shrinking is all the rage nowadays, and despite the demise of the full CCCP, we may yet be optimistic about the prospects for a reduced version of it. The strategy, then, is to retreat to some more restricted formulation of CCCP:

Restricted CCCP: $\qquad P(A \to B) = P(B|A) \quad \text{for all } A, B \in \mathbf{S}$

where $\mathbf{S}$ is some proper subset of F. The version of the Hypothesis that I want to consider first is the following:

Restricted universal version: There is some "→" such that Restricted CCCP holds for all models ⟨ *W, F, P,* → ⟩.

Of course, it is really more a hypothesis schema, because **S** is not properly specified. This will not matter, because any nontrivial version of the restricted universal version must fail. My main purpose is to demonstrate this.

That will mean that with the exception of trivial cases, for any "→," *A*, and *B*, there is a probability function *P* such that $P(A \to B) \neq P(B|A)$. Restricting the propositions that can appear on either side of the "→" simply is not enough. It might be tempting, then, to demand not that Restricted CCCP hold for *all* models but rather that it hold for some privileged subset of them. For example:

Restricted belief function version: There is some "→" such that Restricted CCCP holds for all models ⟨ *W, F, P,* → ⟩ where *P* is a belief function.

Harder though it is to refute this doubly qualified version of the Hypothesis, I shall argue that it, too, is untenable (for any nontrivial restriction).

My attack on these restricted versions of the Hypothesis will begin with an intuitive argument, which I shall then recast in the form of a more rigorous proof. I shall make no assumptions at all about the logic of the "→." I shall demonstrate that, *for a given A and B*, there is *no* relation that *A → B* can bear to *A* and *B* that will allow all probability functions (or even all belief functions, as I shall argue) to conform to the equation.

As I shall show, this sharpens an argument due to Carlstrom and Hill (1978), which in turn is based on an argument by Adams (1975). Appiah (1985) prefers these arguments to those of Lewis (1976), because Lewis assumes that a belief function can be *irregular*, whereas these arguments do not. (An irregular probability function assigns probability zero to other propositions besides the empty one.) The Carlstrom and Hill argument has a small flaw, and some unnecessary, and some unnecessarily strong, assumptions. I believe that my argument corrects the flaw, dispenses with the unnecessary and weakens the unnecessarily strong assumptions, while still avoiding Lewis' irregular probability functions. I also strengthen Lewis' results in other ways.

My proof avoids various other assumptions distinctive of earlier triviality results – assumptions about the closure properties of the set of CCCP-functions for a given "→" (à la Lewis, 1976, 1986), about the cardinality of the models (à la Hájek, 1989), and, as I have said, about the logic of the "→" (à la Stalnaker, 1976). Thus, I believe that triviality results can be had quite cheaply. Just how cheaply is to be gauged by the strength of the one assumption that I do make concerning the uniformity (in a certain sense) of the "→," and I shall argue that even this assumption can be considerably weakened. Where Lewis assumes the uniform interpretation of the "→" throughout certain classes of probability functions [the assumption that Stalnaker (1976) calls "metaphysical realism" and that van Fraassen (1976) disputes], I shall need only some constraints on what non-uniformities there could be.

2. A METHOD FOR GENERATING COUNTEREXAMPLES TO CCCP

The method that I employ is best displayed initially via operations on probability functions that I shall call *perturbations*,[1] but it will be easily seen to encompass other interesting operations, namely, conditionings and Jeffrey conditionings. The trick is similar in each case. We begin by noting a simple home truth. Suppose we have some P, A, and B for which

$$P(A \to B) = P(B|A)$$

Now suppose that some other probability function P' assigns a different probability to the conditional:

$$P'(A \to B) \neq P(A \to B)$$

Then if P' assigns the *same* conditional probability as P does:

$$P'(B|A) = P(B|A)$$

we have immediately

$$P'(A \to B) \neq P'(B|A)$$

Similarly, if P and P' agree on the probability of the conditional, but disagree on the conditional probability, then they cannot both equate the two, for this particular choice of A and B. I shall show that it is easy to find such pairs of probability functions. Now, there are countless ways that two probability functions can be related to one another. I shall show that there are important ways that P and P' can be related that will yield this negative result for the restricted universal version of the Hypothesis.

2.1. Perturbations

The intuitive argument. It is best to present the argument intuitively first. Afterward, I shall provide a more rigorous presentation of it; and after that, I shall derive some philosophical consequences from it. I prefer the intuitive version of the argument to the rigorous one because it not only shows *that* probabilities of conditionals come apart from conditional probabilities but also shows more perspicuously *why* this is so.

I shall picture probability in terms of van Fraassen's "muddy Venn diagram" (1989, p. 161). Propositions are represented by regions on a Venn

116

diagram. Infinitely divisible mud is imagined to be heaped on the diagram, in such a way that the total amount of mud is always taken to be 1, and the amount of mud on any region is just the probability of the proposition represented by that region. Different probability distributions over a fixed algebra are represented by different distributions of mud over the same diagram.

Assume nothing about the logic of "$\rightarrow$," aside from its being a two-place connective. If you like, you may replace "$\rightarrow$" in what follows by some unfamiliar symbol that is free of any associations. Suppose that there is an "$\rightarrow$," a pair of propositions A and B, and a probability function P such that P assigns positive probability to $\neg A$, $A \& B$, and $A \& \neg B$, and $P(A \rightarrow B) = P(B|A)$. P will be our initial distribution of mud. The trick is to find certain movements of the mud that, if allowed, will represent changes in the probability of the conditional, without corresponding changes in the conditional probability, or vice versa.[2] If P' is the distribution of mud that results, then $P'(A \rightarrow B) \neq P'(B|A)$.

Either $A \rightarrow B$ is not a Boolean combination of A and $A \& B$, or it is. [It is important to remember that A and B are not variables here – I am saying that either the unique proposition that is $A \rightarrow B$ is not a Boolean combination of the particular A and $A \& B$ that we are considering, or it is. To say that this particular $A \rightarrow B$ is a Boolean combination of A and $A \& B$ is *not* to say that "$\rightarrow$" is truth functional, because $X \rightarrow Y$ (where X and Y are not A and B) might be some quite different Boolean or even non-Boolean combination of X and $X \& Y$.]

Case 1. $A \rightarrow B$ is not a Boolean combination of A and $A \& B$. Then

(i) $A \rightarrow B$ takes both truth values within $\neg A$, or
(ii) $A \rightarrow B$ takes both truth values within $A \& B$, or
(iii) $A \rightarrow B$ takes both truth values within $A \& \neg B$.

We consider these sub-cases in turn:
(i) $A \rightarrow B$ takes both truth values within $\neg A$:

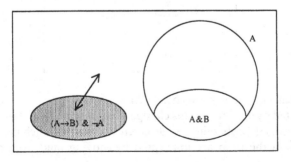

Then we can move mud between $(A \rightarrow B) \& \neg A$ and $\neg (A \rightarrow B) \& \neg A$ (as indicated with an arrow). This changes the total amount of mud on $A \rightarrow B$, that is, it changes the probability of $A \rightarrow B$; however, the conditional probability of B, given A, remains unchanged. After all, $P(B|A) = P(A \& B)/P(A)$, and the terms of this ratio are unaffected by moving mud from one part of $\neg A$ to another. So starting with a probability distribution P for which $P(A \rightarrow B) = P(B|A)$, we can produce others for which this is not so.

(ii) $A \rightarrow B$ takes both truth values within $A \& B$:

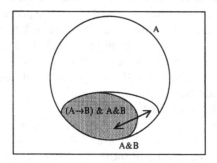

Then we can move mud between $(A \rightarrow B) \& A \& B$ and $\neg (A \rightarrow B) \& A \& B$ (as indicated with an arrow). This changes the probability of $A \rightarrow B$; however, the conditional probability of B, given A, remains unchanged, because the terms of the ratio are unaffected by moving mud from one part of $A \& B$ to another. So starting with a probability distribution P for which $P(A \rightarrow B) = P(B|A)$, we can produce others for which this is not so.

(iii) $A \rightarrow B$ takes both truth values within $A \& \neg B$:

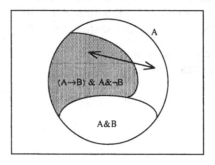

Then we can move mud between $(A \rightarrow B) \& A \& \neg B$ and $\neg (A \rightarrow B) \& A \& \neg B$ (as indicated by an arrow). This changes the probability of $A \rightarrow B$; however, the conditional probability of B, given A, remains unchanged, because the terms of the ratio are unaffected by moving mud from one part of

118

$A \& \neg B$ to another. So starting with a probability distribution P for which $P(A \to B) = P(B|A)$, we can produce others for which this is not so.

Thus we see that if $A \to B$ is not a Boolean combination of A and $A \& B$, we can find functions whose probabilities for $A \to B$ differ from their corresponding conditional probabilities. This completes Case 1.[3]

Case 2. $A \to B$ is a Boolean combination of A and $A \& B$. Because P assigns positive probability to $\neg A$, $A \& B$, and $A \& \neg B$, and $P(A \to B) = P(B|A)$, we can rule out the sub-cases in which $A \to B$ is a tautology or a contradiction. That leaves six sub-cases, which we can group in pairs:

(iv) $A \to B = A \& B$, or $A \to B = \neg (A \& B)$:

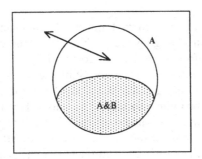

Then moving mud between $\neg A$ and $A \& \neg B$ will change the conditional probability without changing the probability of the conditional. So we can find functions for which the two come apart.

(v) $A \to B = A$, or $A \to B = \neg A$:

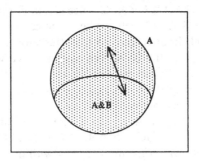

Then moving mud between $A \& B$ and $A \& \neg B$ will change the conditional probability without changing the probability of the conditional. So we can find functions for which the two come apart.

119

(vi) $A \rightarrow B = A \supset B$, or $A \rightarrow B = \neg (A \supset B)$:

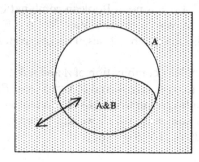

Then moving mud between $\neg A$ and $A \& B$ will change the conditional probability without changing the probability of the conditional. So we can find functions for which the two come apart.

So much for the intuitive argument. Sub-case (iii) is perhaps not a real possibility if the "$\rightarrow$" is a genuine conditional, for it is reasonable to think that $A \rightarrow B$ should be uniformly false throughout $A \& \neg B$. Still, it is worth including this step in the argument, because I do not want it to presuppose anything about the logic of the "$\rightarrow$" – not even something as innocent as this. As a result, the argument embraces even probabilistic or graded "conditionals" such as Lewis discusses (1973, sect. 8), or Halpin's (1991) "$\rightarrow_p$," for which $A \rightarrow_p B$ can be true even when A is true and B false. To repeat, it embraces all two-place connectives, irrespective of their logic.

Thinking again of "$\rightarrow$" as a genuine conditional: Those who favor the principle that $A \& B$ implies $A \rightarrow B$ (as Stalnaker and Lewis do) will deny that sub-case (ii) is a real possibility – they think that $A \rightarrow B$ does not distinguish among the $A \& B$ worlds, since it is uniformly true throughout them. And in general, $X \rightarrow Y$ will not be a Boolean combination of X and $X \& Y$ (for it is well known that no truth function can allow nontrivial conformity to the Hypothesis), so sub-cases (iv)–(vi) will, in general, not obtain (although one of them may obtain for the particular A and B that are being considered). I think the real damage is done in sub-case (i), for I think that it is the normal case.

It may be instructive to note why I had to assume that P assigned non-zero probability to each of $\neg A$, $A \& B$, and $A \& \neg B$. I needed a guarantee that there was some mud to start with in *each* of the three key regions, so that I could make good my threat to move it.

A more rigorous argument. Now, as promised, I shall give a more rigorous reformulation of the foregoing argument. Let $\langle W, F, \rightarrow \rangle$ be given, let P

120

and P' be two probability functions defined thereon, and let $A \in F$ and $B \in F$ be given. Let $A \heartsuit B$ be any one of these combinations of A and B:

(i) $\neg A$
(ii) $A \,\&\, B$
(ii) $A \,\&\, \neg B$

It is a shorthand that will often save me the trouble of writing out all three cases (and you the trouble of reading them), when they can be treated uniformly.

Definition. Call P' a $(\rightarrow, A, B)$-*perturbation of* P if any of the following is the case:

(i–iii) P' and P agree everywhere outside $A \heartsuit B$,[4] but disagree on the probability of $(A \rightarrow B) \,\&\, A \heartsuit B$

(iv) $A \rightarrow B = A \,\&\, B$, or $A \rightarrow B = \neg(A \,\&\, B)$, and $P'(A \,\&\, B) = P(A \,\&\, B)$, but $P'(A) \neq P(A)$

(v) $A \rightarrow B = A$, or $A \rightarrow B = \neg A$, and $P'(A) = P(A)$, but $P'(A \,\&\, B) \neq P(A \,\&\, B)$

(vi) $A \rightarrow B = A \supset B$, or $A \rightarrow B = \neg(A \supset B)$, and $P'(A \supset B) = P(A \supset B)$, but $P'(A \,\&\, B) \neq P(A \,\&\, B)$

Definition. Let A and B be given. Call P *nontrivial relative to A and B* if $P(\neg A)$, $P(A \,\&\, B)$, and $P(A \,\&\, \neg B)$ are all positive.

Limitations of space prevent me from giving here the proofs of the following two lemmas. But really, once you understand the intuitive argument, the proofs practically write themselves. They simply involve a tedious check that the claims hold for each of cases (i)–(vi). [See Hájek (1993) for the details.]

Lemma 1. For all "$\rightarrow$," A, and B, and P nontrivial relative to A and B, there exists a $(\rightarrow, A, B)$-perturbation of P.

Lemma 2. For all "$\rightarrow$," A, and B, if P' is a $(\rightarrow, A, B)$-perturbation of P, then at least one of P' and P violates CCCP for this particular choice of "$\rightarrow$," A, and B.

Definition. Call a proposition A *intermediate* if there is a probability function P such that $0 < P(A) < 1$.

Theorem. There is no "$\rightarrow$," and intermediate, consistent but distinct propositions A and B, such that for all P, $P(A \rightarrow B) = P(B \mid A)$.

Proof. By Lemmas 1 and 2, and the fact that for any such A and B there is a P that is nontrivial relative to them. Q.E.D.

This refutes all (nontrivial) instances of the restricted universal version of the Hypothesis.

It will be handy to have a more general notion of perturbation at the ready. Let "$\rightarrow$" be given.

Definition. Call P' a *perturbation of P relative to* "$\rightarrow$" if there are an A and a B such that P' is a $(\rightarrow, A, B)$-perturbation of P.

I shall sometimes suppress the reference to the "$\rightarrow$," when it is obvious which one is meant. Intuitively, a perturbation of P relative to "$\rightarrow$" is a probability function that, for some A and B, agrees with P everywhere except on some Boolean combination of A and $A \& B$. Their disagreement over that region involves one of the sides of CCCP; their agreement elsewhere ensures their agreement on the other side of CCCP. We could equally speak of perturbation as an *operation*, taking a probability function as input and producing a function that is a specified perturbation of it as output.

As a corollary to Lemma 2, we have what I shall call the Perturbation Theorem:

Perturbation Theorem. If P' is a perturbation of P relative to a given "$\rightarrow$," then at most one of P and P' is a CCCP-function for "$\rightarrow$."

Proof. Suppose that there are an A and a B such that P' is a $(\rightarrow, A, B)$-perturbation of P. Suppose that P is a CCCP-function for "$\rightarrow$." Then $P(A \rightarrow B) = P(B|A)$ for this A and B. But then, by Lemma 2, P' violates CCCP for this particular choice of "$\rightarrow$," A, and B, so P' is not a CCCP-function for "$\rightarrow$." Q.E.D.

As Hájek and Hall (Chapter 6) point out, if we set aside trivial cases, the family of CCCP-functions for a given "$\rightarrow$" is not closed under mixing; and Lewis has shown us that it is not closed under conditioning, conditioning restricted to a finite partition, or Jeffrey conditioning. Now we see that (trivial cases aside) it is not closed under perturbation either. Indeed, we see something stronger still: The family of functions that conform to Restricted CCCP, for a given "$\rightarrow$," is not closed under perturbation (whatever nontrivial restriction we impose). This will have important ramifications for the restricted belief function version, as will be demonstrated shortly.

Comparison with Carlstrom and Hill's argument. Before that demonstration, however, I want to compare my result here with that of Carlstrom and Hill (1978) in their review of Adams – one that Appiah (1985) applauds as being superior to Lewis' (1976) results. Carlstrom and Hill give the following argument, based on an argument by Adams (1975) to show that the probability of a conditional can come apart from the corresponding conditional probability:

Let A and B be contingent (by which I take them to mean "intermediate" in my sense) and logically independent. Assume that $A \to B$ can be true when both A and B are: Let w_1 be a world at which all three are true. Assume that $A \to B$ is not a truth function of A and B. Let w_2 and w_3 be worlds that agree on the truth values of A and $A \& B$, and such that $A \to B$ is false at w_2 but true at w_3. Now consider two probability assignments, P and P', such that P divides almost all the probability roughly equally between w_1 and w_2, and P' divides almost all the probability roughly equally between w_1 and w_3. It follows from this that

$$P(A \to B) \approx \tfrac{1}{2} \quad \text{and} \quad P'(A \to B) \approx 1$$

so

(1)   it is not the case that $P(A \to B) \approx P'(A \to B)$

But

(2) $$P(A \& B) \approx P'(A \& B)$$

and

(3) $$P(A) \approx P'(A)$$

(with both of these positive), and so

(4) $$P(B|A) \approx P'(B|A)$$

Thus, from (1) and (4) we have that at least one of P and P' is a non-CCCP-function for "$\to$."

First, a quibble. The deduction of (4) from (2) and (3) is too quick. Two fractions may have numerators that approximately agree and denominators that approximately agree, without themselves approximately agreeing.[5] To be sure, this small flaw could be patched up – and it is only a quibble, because it is not sufficient to cast doubt on the fact that the probability of the conditional must come apart from the conditional probability for at least one of P and P', and of course it is that which really matters.

More to the point, the argument employs some unnecessarily strong assumptions; correspondingly, the result that is proved is weaker than it could be. Carlstrom and Hill show that a particular assignment of probabilities to worlds leads to trouble for the Hypothesis; but they do not show how restricted the family of CCCP-functions for this "$\to$" must be – and, to be fair, that would have been beyond the scope of their review.

Nevertheless, their argument can be recast in more general form, and then it begins to look like my intuitive argument. Carlstrom and Hill essentially give us an example of a perturbation relative to "$\to$." Start with P's distribution of mud, which heaps roughly half the mud on a

123

world w_2 outside $A \to B$, and move all of this mud (or, more in the spirit of their argument, "roughly" all of it) to a corresponding world w_3 inside $A \to B$ – "corresponding," in the sense that it agrees with w_2 on the truth value of $A \heartsuit B$, where, as before, this stands for one of $\neg A$, $A \& B$, or $A \& \neg B$. This produces P', a particular perturbation of P relative to "$\to$." They prove that two probability functions, related by this specific sort of perturbation, cannot both be CCCP-functions.

The proof lacks generality insofar as it makes what amount to the following assumptions:

A1. $A \to B$ is not a truth function of A and B.
A2. There is a world w_1 at which $A \& B$ and $A \to B$ are true, which gets probability $\frac{1}{2}$ (roughly) from both P and P'.
A3. w_2 is a world inside $\neg(A \to B) \& A \heartsuit B$, w_3 is a world inside $(A \to B) \&$ $A \heartsuit B$, and P assigns probability $\frac{1}{2}$ (roughly) to w_2.
A4. All of w_2's mud (or roughly all of it) is shifted to w_3.

We can simply drop A1: We need no assumptions about the logic of the "$\to$," as I have shown. Furthermore, w_1 actually does no work in the argument, and so A2 should also be dropped – in its place, it suffices simply to assume that P and P' both give A positive probability (to guarantee that the conditional probability does not go undefined). In place of A3, we can assume that there is a nonempty proposition W_2 that implies $\neg(A \to B) \& A \heartsuit B$ and a nonempty proposition W_3 that implies $(A \to B) \& A \heartsuit B$, and P assigns *some* positive probability to W_2 (or to W_3). Finally, in place of A4, we can assume simply that *some* non-zero amount of W_2's mud is moved to W_3 – or vice versa.

Weakening assumptions A1–A4 in the ways that I have indicated, and checking all the cases, we essentially reproduce my argument of Section 2.1. I say "essentially" – I do assume that P and P' agree *exactly* outside $A \heartsuit B$, whereas Carlstrom and Hill assume only that they agree approximately. But this just renders my argument immune to the quibble that I had with their argument – and in any case, the notion of approximate agreement is too vague, as it stands, to be used in a rigorous argument.

A virtue of Carlstrom and Hill's argument over those of Lewis (1976), as Appiah points out, is that it obviates the need for irregular probability functions. In their argument, they concentrate *roughly* all of the probability on certain pairs of worlds – but that leaves a little bit left over that presumably can be spread over the rest of the worlds. Clearly, my result also makes no appeal to irregular probability functions.

Commentary: a plea for perturbation. I said earlier that I would show that various *important* operations on CCCP-functions produce non-CCCP-functions. Why is perturbation important? A perturbation transforms a probability function to another one that is, in a certain sense, similar to it – one that agrees with it everywhere except on some region. (Just how similar the two functions are will depend on how much they differ on this region.) This means that whatever the class of CCCP-functions for a given "→" might be, no two functions in that class are similar in this sense. It also suggests a certain "gappiness" of the CCCP-functions for the "→" among probability functions in general: "Around" each CCCP-function there is a vast cluster of non-CCCP-functions that are otherwise similar to it.

I hope that the notion of perturbation has already proved its worth in enabling the earlier refutation of the restricted universal version of the Hypothesis. However, I think that it has many other uses. Two of the most important of these are canvassed by Hájek and Hall (Chapter 6), and in more detail by Hájek (1993): the doubt that it casts on the Hypothesis expressed in terms of vague probability, as well as on weakened versions of the Hypothesis that replace the exact equality expressed in CCCP by approximate equality. Here I want to discuss some other ramifications of the Perturbation Theorem: the argument that it provides against the restricted belief function version of the Hypothesis, its application to sensitivity analysis, perturbations of the boundaries of propositions, and the unification that it provides for Lewis' triviality results.

An argument against the restricted belief function version of the Hypothesis. Hájek and Hall (Chapter 6) show the belief function version to be untenable; in my opinion, the Perturbation Theorem spells the demise of the restricted belief function version also (for any nontrivial restriction). To uphold it, no two probability functions that represent the states of mind in the epistemic history of a rational agent can be perturbations of each other. Nor can the probability functions of two different rational agents be perturbations of each other relative to some "→." To save the restricted belief function version, we need to be convinced that at most one of any two probability functions related by perturbation can be a belief function. But this will be no easy task. A perturbation of a probability function can differ ever so slightly from that function – for example, agreeing with it virtually "everywhere", and elsewhere disagreeing with it only beyond the hundredth decimal place. It is hard to see how one of these functions can be a belief function, and the other not.

The friend of the Hypothesis may insist that we follow Lewis' suggestion that "the content of a total mental state is the system of belief and desire that best rationalizes the behavior to which that state would tend to dispose one" (1986, p. 585), and that while a function that does not conform

125

to Restricted CCCP may fit an agent's mental state, it cannot *rationalize* it.[6] (After all, he may remind us, the restricted belief function version does see Restricted CCCP as a constraint on rationality.) Thus it is no embarrassment for the Hypothesis that one function is a belief function, whereas a very near neighbor of it is not – the extents to which they fit a certain agent's mental state are roughly the same, but whereas one rationalizes the behavior to which that state would tend to dispose one, the other does not. Or so the argument goes.

This sort of move is always open, although it is one that should raise the hackles of anyone with Popperian sympathies – little surprise that CCCP holds throughout the class of all belief functions (on an appropriately restricted domain), because any function that fails to conform to it is thereby disqualified from being a belief function. In the absence of good arguments to support this, it seems that the restricted belief function version is being rescued, as the old joke goes, simply by stipulating that there are no counterexamples to it!

In any case, functions related by perturbation do not *have to* be very near neighbors. For example, they may agree that A should be given a certain very tiny probability, but disagree wildly on how probability should be distributed within $\neg A$. We will need to be convinced once again that at most one of those functions is a belief function, and, frankly, I do not see how that can be done in a non-question-begging way, without an appeal to the Hypothesis itself; for I see no reason to doubt that two such functions can be belief functions, apart from a prior faith in the Hypothesis. To be sure, there may be a *consistent* package deal of the Hypothesis plus elevated standards of what it takes to be a belief function. But consistency is no great virtue of a position (we learned from Quine that one can hang on to virtually any cherished belief if one is prepared to twist and turn enough elsewhere). What this position lacks, in my opinion, is *plausibility*.

Robustness, sensitivity analysis.[7] Let us move on to other applications of the Perturbation Theorem. Following Walley (1991, p. 5), for example, we can call the conclusion of a statistical analysis *robust* if that same conclusion is reached by the use of each member of a wide class of probability functions. Those who are uncomfortable about the Achilles' heel of Bayesianism, that the choice of priors is arbitrary, but who are otherwise sympathetic to the Bayesian enterprise, may find robustness to be a useful notion. The robustness of an inference or decision can be checked by performing a so-called sensitivity analysis. Take a large class of priors, and derive posterior distributions from each of them (by conditioning, say, or by a specified sequence of conditionings). If the same inference or decision is reached on the basis of all of these posterior distributions, then it is robust:

The choice of prior was not critical. [See Walley (1991) for further discussion and references.]

Suppose that included among the modifications of a given prior are perturbations of that prior (relative to some "→"). Although a specific choice of prior may, for all we have learned so far, be a CCCP-function for that "→," these modifications will not be. Conformity to CCCP forces one to be very careful that the prior, and its modifications, are just right.

Perturbations of the boundaries of propositions. Here is another use for perturbations. Keeping in mind the muddy Venn diagram, notice that perturbations involve movement of mud *relative to the boundaries of propositions.* I assumed earlier that the boundaries of the propositions were fixed and that it was the mud that moved, but the same effect could be achieved by keeping the distribution of mud fixed and moving the boundaries, so to speak. Here's one way that this might come about: Suppose that the "→" depends partly on a similarity relation on worlds. Imagine an agent who changes her mind about the similarity relation, without actually learning anything about the world in which she lives. She might, for example, read Lewis' *Counterfactuals* and become persuaded that sameness of laws is a more important respect of similarity of worlds than she previously thought, or she might be in some context in which a similarity relation that is finer-grained than before becomes appropriate. In neither case does she learn anything about *her* world – her assignment of probabilities to individual worlds is as it was before. But for at least some A and B, $A \to B$ will be a different set of worlds from what it was previously. This we might picturesquely call *a revision in the boundary of* $A \to B$ (after all, that is how we would picture it on the muddy Venn diagram).

Certain revisions in the boundaries of propositions of this sort would have the same effect as perturbations of her old probability assignment. We could even extend our usage of the word and call such revisions "perturbations" of the propositions concerned. For example, a perturbation of $(A \to B) \& \neg A$ involves a revision in the boundary of that proposition, such that the new assignment of probabilities to propositions is a type-(i) $(\to, A, B)$-perturbation of the old. Note that there is always a conditional that is involved in a perturbation – and it is propositions involving conditionals, above all, that we should reasonably expect to undergo occasional revision in the course of an agent's deliberation (because their truth depends partly on what the agent takes the similarity relation to be, something that is not fixed simply by the facts about the worlds themselves).

There is yet another way to achieve the same effect of relative mud movement: move *both* the mud *and* the boundaries of propositions at the same time. Indeed, someone who believes that the conditional is radically dependent on the probability function, changing with every change of mind of the agent who entertains it, believes that this is how things really are. Van Fraassen (1976) endorses this view in his critique of Lewis' triviality results. Again, certain of these joint movements will have the same effects as perturbations of probability functions (and we could express their kinship by giving them the same name). I shall discuss them again at the end of this chapter.

Unifying Lewis' triviality results. There is an interesting connection between conditioning and perturbation. Consider the special case of perturbation in which all of the mud, x, say, is removed from a region W that implies $\neg (A \to B) \& A \heartsuit B$; and then all of it is deposited in some way on a region W' that implies $(A \to B) \& A \heartsuit B$ and that previously had no mud on it. (This could also be done in the reverse direction, with suitable changes in the discussion that follows.) Let P be the distribution before the amount x of mud is shifted, and P' the distribution afterward. Then there is an ur-distribution P_0 such that both P and P' are derived from P_0 by conditioning. Here's the idea: P_0 is a certain distribution that gives equal probability to W and W'. Punch a hole through W' and renormalize, and you have P; punch a hole through W and renormalize, and you have P'.[8] Precisely: P_0 is the distribution that gives the same odds as P and P' do to all propositions that are incompatible with both W and W', the same odds to propositions that imply W as P does, and the same odds to propositions that imply W' as P' does, but it gives probability $x/(x + 1)$ to both W and W'. P is derived from P_0 by conditioning on $\neg W'$; P' is derived from P_0 by conditioning on $\neg W$.

Suppose that the set of CCCP-functions for some "$\to$" contains P_0. Then we have found that it is not closed under conditioning – for it does not contain both P and P'. Now suppose, instead, that the set *is* closed under conditioning, as Lewis (1976) does in order to derive his second triviality result. Then we know that it does not contain any probability function that gives the same non-zero probability to a proposition inside $(A \to B) \& A \heartsuit B$ as it does to a proposition inside $\neg (A \to B) \& A \heartsuit B$ – for if it did, we could take that function to play the role of P_0 and use the foregoing argument and the Perturbation Theorem to show that one can derive a non-CCCP-function from it by conditioning, *contra* the closure assumption.

The trouble is that we have good reasons for thinking that the set of CCCP-functions for a given "$\to$" *does* contain a probability function that gives the same non-zero probability to some proposition inside $(A \to B) \&$ $A \heartsuit B$ as it does to some proposition inside $\neg (A \to B) \& A \heartsuit B$, for some

A, B, and ♥. Indeed, this would seem to be the norm for CCCP-functions, because it is known that they have very rich domains. Hájek (1989) shows that, without any assumptions about the logic of the "$\rightarrow$," all CCCP-functions have infinite domains; and Hall (Chapter 8, this volume) shows that, with minimal assumptions about the logic, all CCCP-functions have uncountable domains – indeed, that any probability space whose probability function is a CCCP-function is *full*. $\{\langle W, F, P, \rightarrow \rangle$ is full if, for each element A of F, P takes every value in $[0, P(A)]$ on the elements of F that are subsets of A.$\}$

So let P_0 be a CCCP-function that gives positive probability to both $(A \rightarrow B) \& A$ ♥ B and $\neg(A \rightarrow B) \& A$ ♥ B, for some A, B, and ♥ – it would surely trivialize the set of CCCP-functions for this "$\rightarrow$" if there were no such function. Because its probability space is full (making Hall's assumption about the "$\rightarrow$"'s logic), there is a proposition W that implies $\neg(A \rightarrow B) \&$ A ♥ B, and a proposition W' that implies $(A \rightarrow B) \& A$ ♥ B, such that $P_0(W) = P_0(W') > 0.$[9] However, $P = P_0(- | \neg W')$ and $P' = P_0(- | \neg W)$ are perturbations of each other, so they cannot both be CCCP-functions. Thus, the set of CCCP-functions for this "$\rightarrow$" is not closed under conditioning. Our study of perturbations has enabled us to produce what is basically a new proof of Lewis' second triviality result.[10]

Conditioning is a special case of Jeffrey conditioning, which more generally involves a change in the probability weights given to the members of some finite partition, while preserving the odds of all propositions within each of the cells of the partition. Various perturbations are Jeffrey conditionings. Of course, any perturbation for a model that has finitely many worlds is a Jeffrey conditioning, for the cheap reason that *any* mud movement whatsoever in such a model is: Take the partition to consist of the individual worlds themselves, so that any mud movement will involve such a reassignment of weights to the members of the partition. Similarly, in models with infinitely many worlds, there are also perturbations that are trivially Jeffrey conditionings: for example, a perturbation involving only finitely many worlds of that model, with the partition consisting of those worlds, and the union of everything else.

But there are less trivial examples of perturbations that are Jeffrey conditionings. Suppose that for some A, B, and ♥, and for some W that implies $\neg(A \rightarrow B) \& A$ ♥ B, and some W' that implies $(A \rightarrow B) \& A$ ♥ B, we shift an amount of mud from W to W', in such a way that the odds of all propositions that imply W and that imply W' remain the same. This is a perturbation that is a Jeffrey conditioning over the partition $\{W, W', \neg(W \vee W')\}$, with the weights of only the first two cells changing. Such perturbations are useful in modeling belief change that is not global (the way that conditioning is), but local – when an agent learns something that gives probabilistic information about a specific region of logic space, without this information having effects that spread across the whole space.

And we can generalize further. Consider an m-celled partition of $(A \to B) \& (A \heartsuit B)$, giving us an $(m + 2)$-celled partition of the whole space – the $(m + 1)$st cell being $\neg (A \to B) \& (A \heartsuit B)$, and the $(m + 2)$nd cell being $\neg (A \heartsuit B)$. Suppose we shift various amounts of mud from cells 1 through m, to cell $m + 1$ (as we can in appropriately nontrivial cases), in such a way that the odds of all propositions within these cells remain the same. This is a Jeffrey conditioning over the $(m + 2)$-celled partition; the same is true if we move the mud in the opposite direction. And again, because it is a perturbation, if the prior function (before the shift) is a CCCP-function for the "$\to$," the posterior function (after the shift) is not.

Call an instance of Jeffrey conditioning n-celled if it takes place over an n-celled partition. I have thus proved the following corollary to Lewis' fourth triviality result: There is no CCCP-conditional for a class of probability functions closed under n-celled Jeffrey conditioning, for each $n \geqslant 3$, unless that class consists entirely of trivial functions.

I have said enough about perturbations – the illustration they provide of a general method for refuting the restricted universal version of the Hypothesis, their role in generalizing the Carlstrom and Hill argument, their interest in their own right, and their relations to important forms of belief revision. I promised that the method for refuting the restricted universal version encompasses other interesting operations on probability functions. Let us see how.

2.2. Conditioning[11]

I began my discussion of conditioning in the preceding section, and it is now time to say more. Start with propositions A, B, and $A \to B$. Suppose P is a CCCP-function for this "$\to$," and conditionalize it on a proposition E, arriving at a new probability function P_E. Then there are many choices of E such that $P_E(A \to B) \neq P(A \to B)$, and yet $P_E(B|A) = P(B|A)$. Thus, P_E will be a non-CCCP-function – and indeed it will violate the equation in CCCP for this very choice of $A \to B$.

1. Here is a simple example of such an E:

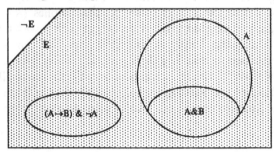

E is implied by $(A \to B) \vee A$, with $0 < P(E) < 1$. [It is not essential that $(A \to B) \& \neg A$ be nonempty – the diagram may lack the ellipse.] Conditioning on E amounts to what we might call "trimming" the original diagram and renormalizing what remains so that it once again has a total of one unit of mud. The probability of $A \to B$ after conditioning is greater than it was before, because $A \to B$ receives a greater proportion of the new distribution of the mud than it did of the old. But the corresponding conditional probability is unchanged by the conditioning. This is obvious enough when it is noted that the proportion of A mud that falls inside $A \& B$ remains unchanged, but I should prove it:

Suppose that E is implied by $(A \to B) \vee A$, that $P(E) < 1$, and that $P(A \to B) = P(B|A)$. Then

$$P_E(A \to B) = P(A \to B|E)$$

$$= \frac{P(A \to B)}{P(E)} \quad \text{(because } A \to B \text{ implies } E)$$

$$> P(A \to B) \quad \text{[because } P(E) < 1]$$

Thus $P_E(A \to B) > P(A \to B)$. But

$$P_E(B|A) = \frac{P_E(A \& B)}{P_E(A)}$$

$$= \frac{P(A \& B|E)}{P(A|E)}$$

$$= \frac{P(A \& B \& E)}{P(A \& E)}$$

$$= \frac{P(A \& B)}{P(A)} \quad \text{(because } A \& B \text{ and } A \text{ both imply } E)$$

$$= P(B|A)$$

Thus $P_E(B|A) = P(B|A)$, and hence $P_E(A \to B) > P_E(B|A)$.

Now that we have seen the trick, it is easy to come up with other examples of propositions on which a CCCP-function cannot be conditionalized without producing a non-CCCP-function – indeed, one that violates the equation in CCCP for this very choice of $A \to B$. (I assume that these propositions all have positive probability, so that they *can* be conditionalized on.) I do not see any need to give the proofs for the next examples – they are equally straightforward and can be easily read off the diagrams:

2. Let E be any proposition whose probability is less than 1, and whose negation implies

(i) $\neg (A \to B) \& \neg A$,
(ii) $\neg (A \to B) \& A \& B$,

or the conjunction of any such propositions. These are like "hole-punches":

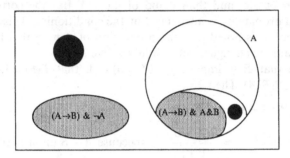

3. Let E be any proposition whose probability is less than 1, and whose negation implies

(i) $(A \to B) \& \neg A$,
(ii) $(A \to B) \& A \& \neg B$,

or the conjunction of any such propositions. More hole-punches:

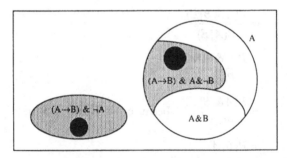

It is worth noting that Cases 2(ii) and 3(ii) – if such there be – do not exemplify the strategy of increasing or decreasing one side of CCCP while keeping the other constant. Rather, they are examples of something *worse still*, as far as CCCP is concerned. Conditioning on a proposition whose negation implies $\neg (A \to B) \& A \& B$ increases the probability of the conditional, while decreasing the conditional probability; conditioning on a proposition whose negation implies $(A \to B) \& A \& \neg B$ decreases the former, while increasing the latter. I say "if such there be" because Case 2(ii) is ruled out if $A \& B$ implies $A \to B$; and, very plausibly, $A \& \neg B$ implies

132

$\neg (A \rightarrow B)$, and this rules out Case 3(ii). The other cases, however, are unproblematic. Apart from cases that we can justly call trivial, given "$\rightarrow$," A, and B, such examples of E can always be found.

We see that for any "$\rightarrow$," the family of CCCP-functions is not closed under conditioning (apart from trivial cases). In fact, we see something stronger: For any "$\rightarrow$," A, and B, the set of functions for which the probability for $A \rightarrow B$ equals the corresponding conditional probability is not closed under conditioning (apart from trivial cases). Thus, we have strengthened Lewis' second triviality result:

Theorem. There is no "$\rightarrow$" such that Restricted CCCP holds throughout a class of probability functions closed under conditioning, irrespective of the (nontrivial) restriction that we impose, unless the class consists entirely of trivial functions.

It appears that conditioning is a dicey business if you are an agent who wants to conform to CCCP. You must be careful that for any E that you conditionalize on, there is *no* pair of propositions such that E falls into one of the outlawed categories with respect to them. This might have seemed manageable if your probability space had not been very rich (so that there would not be that many pairs of propositions to worry about). But your probability space *has to be* very rich – certainly at least infinite, and plausibly even uncountable, as I've noted.

These are examples of propositions fit for Moore's paradox – ones that you regard as possible, but on which you nevertheless cannot conditionalize without violating CCCP. Van Fraassen (1984) calls propositions on which one cannot conditionalize without violating such a structural constraint "Moore propositions." Regarding conditioning as the model of ideal learning, this means that the ideally rational agent cannot learn such propositions while adhering to the constraint. That a certain proposition should be a Moore proposition is not surprising if it is "reflexive" in some appropriate sense (which I shall not try to spell out here) – something about the agent's own beliefs, say, or about a future history that a theory of chance says would undermine that very theory (Lewis, 1981). But what is disturbing for the Hypothesis is that the preservation of CCCP turns so *many* propositions into Moore propositions, seemingly without explanation.

In one sense, this is overkill. Hájek and Hall (Chapter 6) show that the result of nontrivially conditioning a CCCP-function on some proposition is always a non-CCCP-function. So the existence of these various Moore propositions follows immediately.

But it is overkill only *in one sense*. First, I think that the method here gives us a way of seeing intuitively *why* the preservation of CCCP turns various propositions into Moore propositions, whereas that may not be so obvious from the proof in Chapter 6, which is more algebraic. Second, and more important, the other result assumes that the "→" is uniform across a pair of probability functions: Its interpretation is the same whether it appears within the scope of a certain probability function or another one reached by conditioning that function. The argument here can go through without this assumption, as will be demonstrated in Section 3.

2.3. Jeffrey conditioning

Conditioning is a special case of Jeffrey conditioning, as I have said; so the examples in Section 2.2 could equally be described as Jeffrey conditionings that take a CCCP-function to a non-CCCP-function. In fact, there are many other Jeffrey conditionings with this property. We saw some of them in Section 2.1, and I shall mention a few more here.

To get the idea, consider again the case of conditioning on an E that "trimmed" the diagram. We removed all the $\neg E$ mud and renormalized – or, equivalently, redistributed the original mud to E – preserving the odds of all propositions that imply E. Now suppose that we "shave" the diagram instead, moving some but not all of the $\neg E$ mud to E, in such a way that the odds of all propositions that imply $\neg E$, and of all propositions that imply E, are preserved. This is a Jeffrey conditioning, on the partition $\{\neg E, E\}$. It is also incompatible with the preservation of CCCP, for this mud shift increases the probability of $A \to B$, without increasing the corres-ponding conditional probability. (Obviously, a Jeffrey conditioning in the other direction, which would decrease the probability of $A \to B$, would equally violate CCCP.)

More precisely, let "→," A, and B be given, let P be a CCCP-function for this "→," let E be implied by $(A \to B) \lor A$, and let P' be a function derived from P by a Jeffrey conditioning on the partition $\{\neg E, E\}$ that increases the probability of E. We have

$$P'(A \to B) = P(A \to B \mid E)P'(E) + P(A \to B \mid \neg E)P'(\neg E)$$

$$= \frac{P(A \to B)}{P(E)} P'(E) \quad \text{(because } A \to B \text{ implies } E)$$

$$> P(A \to B) \quad \text{[because, by hypothesis, } P'(E) > P(E)]$$

Hence

(5) $$P'(A \to B) > P(A \to B)$$

134

However,

$$P'(B|A) = \frac{P'(A \& B)}{P'(A)}$$

$$= \frac{P(A \& B|E)P'(E) + P(A \& B|\neg E)P'(\neg E)}{P(A|E)P'(E) + P(A|\neg E)P'(\neg E)}$$

$$= \frac{[P(A \& B)/P(E)]P'(E)}{[P(A)/P(E)]P'(E)} \quad \text{(because } A, \text{ and hence } A \& B, \text{ implies } E\text{)}$$

$$= \frac{P(A \& B)}{P(A)}$$

$$= P(B|A)$$

Hence

$$P'(B|A) = P(B|A)$$

and so, by (5), and the fact that P is a CCCP-function for "$\rightarrow$,"

$$P'(A \rightarrow B) > P'(B|A)$$

So such a Jeffrey conditioning is incompatible with the preservation of CCCP, and indeed any function so produced violates the equation in CCCP for this very choice of A and B.

Clearly, any of the Moore propositions listed in Section 2.2 will play the role of such an E here – giving us various $\{\neg E, E\}$ partitions on which we cannot Jeffrey-conditionalize.

This, in turn, implies that for any "$\rightarrow$," A, and B, the class of functions for which the probability for $A \rightarrow B$ equals the corresponding conditional probability is not closed under two-celled Jeffrey conditioning (apart from trivial cases). But it follows from this that it is not closed under n-celled Jeffrey conditioning for each $n \geqslant 2$ (for any two-celled Jeffrey conditioning can be mimicked by an n-celled Jeffrey conditioning, in which $n - 1$ cells are given the same weight, and thus jointly play the role of one of the two original cells). Thus, we have the following strengthening of Lewis' fourth triviality result:

Theorem. For each n, there is no "$\rightarrow$" such that Restricted CCCP holds throughout a class of probability functions closed under n-celled Jeffrey conditioning, irrespective of the (nontrivial) restriction that we impose, unless the class consists entirely of trivial functions.

135

3. Assessment: a loophole?

We have found that, assuming nothing about the logic of the conditional, all perturbations, some conditionings, and some (further) Jeffrey conditionings performed on a CCCP-function produce non-CCCP-functions. Moreover, specify any particular A, B, and "$\rightarrow$" that you like, and a probability function P for which $P(A \rightarrow B) = P(B|A)$, and we can produce a P' by application of any of these operations on P, such that P' violates the equation in CCCP for this very choice of A, B, and "$\rightarrow$."

Let us return to the argument of Section 2.1, which is a good example of the method that I have employed – the intuitive argument involving perturbation (or, if you prefer, its more rigorous reformulation). I have discussed how it generalizes the Carlstrom and Hill result, and how it can be used to unify the results of Lewis. Now I want to draw further comparisons to the latter. Like Lewis, I assumed nothing about the logic of "$\rightarrow$." Lewis showed that any class of CCCP-functions closed under conditioning (restricted conditioning, Jeffrey conditioning) consists entirely of trivial functions. I derived analogous results – for example, that any class of CCCP-functions closed under perturbation consists entirely of trivial functions. However, as Hájek and Hall (Chapter 6) show, Lewis' results can be strengthened: Nontrivially conditioning a CCCP-function always produces a non-CCCP-function. And this parallels my result too (replacing "nontrivially conditioning" by "perturbing").

The assumption of Lewis that has received the most criticism is the fixed interpretation of the "$\rightarrow$" – the assumption that Stalnaker calls "metaphysical realism." It may be thought – and van Fraassen (1976) does think – that the "$\rightarrow$" is inextricably tied to the probability function in whose scope it appears: Change the function, and you change the "$\rightarrow$." Moreover, if such radical context dependence of the "$\rightarrow$" is embraced, a positive result due to van Fraassen shows that a certain version of the Hypothesis is actually tenable: Roughly, for every full probability space $\langle W, F, P \rangle$, an "$\rightarrow$" can be tailored such that $\langle W, F, P, \rightarrow \rangle$ is a model for which CCCP holds. (See Hájek and Hall, Chapter 6, for further details and qualifications.) This is the sense in which the Hypothesis is "immortal," in the words of my opening paragraph. So radical context dependence is an antidote to even the strengthening of Lewis' results – not an antidote that I can swallow (for reasons given in Chapter 6), but others can.

It might seem that radical context dependence is also an antidote to my intuitive result from perturbation – and hence to the results that follow it. Van Fraassen's result shows us that, to a certain extent, this must be right. Let us see how context dependence can disarm my intuitive argument, by considering the following diagram.

136

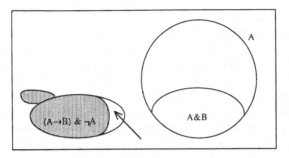

Dodging the mud.

Suppose that some mud that currently lies outside $(A \to B)\& \neg A$ is to move to certain worlds that are currently inside $(A \to B)\& \neg A$. A perturbation threatens. However, it can be prevented if the boundary of $(A \to B)\& \neg A$ moves too, in just such a way as to "dodge" the approaching mud.

But not just any movement of the boundary will do. The probability of $A \to B$ must not change in any way, despite the movement – that would be equally unpropitious for the Hypothesis, for the conditional probability still remains unchanged. So while the boundary retreats to avoid the scene of the mud movement, it must compensate elsewhere, invading territory that it did not occupy previously.

Alternatively, $A \to B$ could "accept" the moved mud provided that its boundary shrank so as to keep the total amount of mud inside constant:

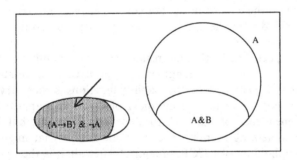

Accepting the mud.

Finally, there are intermediate cases, in which $A \to B$ changes so as to dodge some of the moved mud, while accepting the rest. (In these cases, not all of the mud is moved to a single world.)

And so it is with the various other forms of perturbation: If the boundary of the conditional always moves in just the right ways, in tandem with

137

the movement of mud – which is to say that the conditional is always reinterpreted in just the right ways, in tandem with the change in probability distribution – then all is well for the Hypothesis. However, this should not necessarily provide the friend of the Hypothesis much solace. As I noted in Section 2.1, the effect of a perturbation can be achieved *even if* the "→" is radically context-dependent – a joint mud/boundary shift can still be trouble. In fact, certain boundary shifts only aggravate the effect of a given mud movement. (A shift that invades new territory *and* accepts all of the moved mud is an example of this.) And certain other boundary shifts, though not aggravating the effect of a given mud movement, do not counterbalance it enough. The only hope for the upholding of CCCP is if the boundary shifts are always of just the right kind.

This is as true for the cases of conditioning and Jeffrey conditioning as it is for the cases of perturbation. Remember the case of "trimming" (or "shaving") the diagram, which raises the probability of $A \rightarrow B$, without raising the corresponding conditional probability. Suppose that the "→" also changes with the change in probability distribution, in such a way that the new $A \rightarrow B$ includes all the worlds that it did previously, *plus some new worlds* that receive non-zero probability. This is doubly bad news for CCCP: $A \rightarrow B$ is raised both by the effect of the (Jeffrey) conditioning *and* by the growth in its boundary, while the conditional probability remains unchanged. In a similar vein, my argument can allow the boundary of $A \rightarrow B$ to change simultaneously with a conditioning on any of the other Moore propositions catalogued in Section 2.2, if the change accentuates the deleterious effect (as far as CCCP is concerned) of the conditioning; or the effect of the change in the boundary can even counteract the effect of the change due to the conditioning, provided the two do not exactly cancel.

In this respect, I think that my result buys triviality more cheaply than Lewis' results, even after strengthening: The fixed interpretation of the "→" across (at least) a pair of probability functions is necessary for those results, but not for mine. The upholder of CCCP can appeal to radical context dependence of the "→" to elude Lewis' results, but he has to appeal to something even stronger to elude mine. What might that be? Why, the upholding of CCCP, of course! After all, it is the upholding of CCCP that would guarantee that the "→" is sensitive to the probability function in just such a way that CCCP is upheld. But I do not see how *else* he could justify the assumption that the "→" not only depends on the probability function but also depends on the probability function in just the right way. I concede once again that his position is consistent – van Fraassen's positive result assures us of that – but again, *that* is scant praise for it.[12]

138

Notes

1. This term was suggested to me by David Lewis.
2. During a discussion of our joint contribution (Chapter 6, this volume), Ned Hall pointed out to me that mud movements over the $\neg A \& (A \to B)$ boundary have this property; he is thus an important source of inspiration for my result.
3. Actually, there are other mud movements that have the same effect. For instance, one could do two or three of the mud movements described in (i)–(iii) at once, provided there is a net change of mud in $A \to B$.
4. In case it is not already obvious what this means: For all propositions $X \in F$ that imply $\neg (A \heartsuit B)$, $P(X) = P'(X)$.
5. For example, if we read "$x \approx y$" as "differs by no more than ε from," for some suitably small $\varepsilon > 0$ that has been chosen, then it is clear that we can have (2) and (3) without (4): Simply let $P(A \& B) = P'(A \& B) + \varepsilon$, and $P(A) = P'(A) - \varepsilon$, so that the numerators of the conditional probabilities differ, but in the opposite way to the denominators, thus exacerbating the difference in their ratios.
6. Lewis offers the friend of regularity such a response in the face of his third triviality result.
7. I thank Bas van Fraassen for drawing this application to my attention.
8. It was David Lewis who pointed out (a close relative of) this fact to me.
9. In case this is not already obvious: Suppose $P_0(\neg (A \to B) \& A \heartsuit B) \leqslant P_0((A \to B) \& A \heartsuit B)$. Then take W to be $\neg (A \to B) \& A \heartsuit B$; by the fullness of the space, we are guaranteed the existence of an appropriate W'. We reason similarly if the inequality goes the other way.
10. Only "basically," because "triviality" in this case does not mean the same thing as it does in Lewis' result. The ensuing comparisons to Lewis' triviality results should be qualified similarly.
11. I thank David Lewis and Lyle Zynda for helpful discussion on this section.
12. I thank Ned Hall, David Lewis, Brian Skyrms, Mike Thau, Bas van Fraassen, Jim Woodward, and Lyle Zynda for valuable discussion. I am also grateful to the Whiting Foundation for its financial support during this project.

References

Adams, Ernest (1975) *The Logic of Conditionals.* D. Reidel, Dordrecht.

Appiah, Anthony (1985) *Assertion and Conditionals.* Cambridge University Press.

Carlstrom, Ian F., & Hill, Christopher S. (1978) "Review of Adams' *The Logic of Conditionals.*" *Philosophy of Science* 45: 155–8.

Hájek, Alan (1989) "Probabilities of Conditionals – Revisited." *Journal of Philosophical Logic* 18: 423–8.

Hájek, Alan (1993) "The Conditional Construal of Conditional Probability." Ph.D. dissertation, Princeton University.

Halpin, John (1991) "What Is the Logical Form of Probability Assignment in Quantum Mechanics?" *Philosophy of Science* 58: 36–60.

Harper, W. L., Stalnaker, R., & Pearce, G. (eds.) (1981) *Ifs.* D. Reidel, Dordrecht.

Lewis, David (1973) "Counterfactuals and Comparative Possibility." *Journal of Philosophical Logic* 2: 418–46. Reprinted in Harper et al. (eds.) (1981) *Ifs.* D. Reidel, Dordrecht.

Lewis, David (1976) "Probabilities of Conditionals and Conditional Probabilities." *Philosophical Review* 85: 297–315; reprinted in Harper et al. (eds.) (1981) *Ifs.* D. Reidel, Dordrecht.

Lewis, David (1981) "A Subjectivist's Guide to Objective Chance." Pp. 267–97 in *Ifs*, ed. W. L. Harper, R. Stalnaker, & G. Pearce. D. Reidel, Dordrecht.

Lewis, David (1986) "Probabilities of Conditionals and Conditional Probabilities II." *Philosophical Review* 95: 581–9.

Lewis, David (1992): "Humean Supervenience Debugged." Presented at the 1992 Australasian Association of Philosophy Conference, Brisbane, Australia.

Stalnaker, Robert (1970) "Probability and Conditionals." *Philosophy of Science* 37: 64–80; reprinted in Harper et al. (eds.) (1981) *Ifs*. D. Reidel, Dordrecht.

Stalnaker, Robert (1976) "Letter to van Fraassen." Pp. 302–6 in *Foundations of Probability Theory, Statistical Inference, and Statistical Theories of Science*, vol. I, ed. W. L. Harper & C. A. Hooker. D. Reidel, Dordrecht.

van Fraassen, Bas (1976) "Probabilities of Conditionals." Pp. 261–301 in *Foundations of Probability Theory, Statistical Inference, and Statistical Theories of Science*, vol. I, ed. W. L. Harper & C. A. Hooker. D. Reidel, Dordrecht.

van Fraassen, Bas (1984) "Belief and the Will." *Journal of Philosophy* 81: 235–56.

van Fraassen, Bas (1989) *Laws and Symmetry*. Clarendon Press, Oxford.

Walley, Peter (1991) *Statistical Reasoning with Imprecise Probabilities*. Chapman & Hall, London.

8

Back in the CCCP

NED HALL

By the "conditional construal of conditional probability" (CCCP) I mean the claim that conditional probability is the probability of some conditional. Over the past two decades, many authors have scrutinized this claim, amassing a number of results about its tenability – mostly negative. In this chapter I establish three new results, also negative. Is this overkill? Perhaps, but the theorems I prove are, on balance, much more general and logically quite a bit stronger than any other results of which I am aware. (For example, David Lewis' four triviality results fall out as special cases of the first of my theorems.) At any rate, I leave aside discussion of the significance of this work for CCCP's tenability, because this is intended as a technical treatment; philosophical evaluation of CCCP and the technical literature to which it has given rise is best left for another work.[1] However, after describing and proving my results, I shall offer some comments designed to illuminate their relation to the other significant results. But first, some preliminaries.

PRELIMINARIES

Call a triple $\langle W, \mathsf{F}, P \rangle$ a *probability space* iff W is a set, F a set of subsets of W, and P a countably additive probability function defined on the elements of F. (Heuristic: W is a set of possible worlds; F is the set of propositions over which P is defined.) We require that F be closed under the usual Boolean operations of intersection, union, and complement within W, and that F be a sigma-algebra – any countable union of members of F is a member of F.

Call a quadruple $\langle W, \mathsf{F}, P, \rightarrow \rangle$ a *model* iff $\langle W, \mathsf{F}, P \rangle$ is a probability space and "$\rightarrow$" is a binary function with domain F and range a subset of F: For any $A, B \in \mathsf{F}$, $A \rightarrow B \in \mathsf{F}$. Similarly, call the triple $\langle W, \mathsf{F}, \rightarrow \rangle$ an *algebra*.

CCCP holds for a model $\langle W, \mathsf{F}, P, \rightarrow \rangle$ iff for all A, B such that $P(A) \neq 0, P(B|A) = P(A \rightarrow B)$. Given an algebra $\langle W, \mathsf{F} \rightarrow \rangle$, call a function

141

P defined on F a *CCCP-function* for the algebra iff CCCP holds for the model $\langle W, F, P \rightarrow \rangle$.

A probability space $\langle W, F, P \rangle$ or model $\langle W, F, P, \rightarrow \rangle$ is *trivial* iff P has, at most, four distinct conditional probability values.[2] Negative results about CCCP stop at triviality, because it is always possible to extend a trivial probability space to an equally trivial model for which CCCP holds.

Given a model $\langle W, F, P, \rightarrow \rangle$, a proposition A is a *P-atom* iff for all B, either $P(AB) = 0$ or $P(AB) = P(A) > 0$.

Two models $\langle W_1, F_1, P_1, \rightarrow_1 \rangle$ and $\langle W_2, F_2, P_2, \rightarrow_2 \rangle$ *employ the same arrow* iff the corresponding algebras $\langle W_1, F_1, \rightarrow_1 \rangle$ and $\langle W_2, F_2, \rightarrow_2 \rangle$ are isomorphic.

When two models employ the same arrow, we can, without loss of generality, simply identify their various constituents (except for the probability functions, of course) and go on to describe them as consisting of different probability functions defined on the same algebra.

Call two probability functions P and P' defined on the same algebra $\langle W, F, \rightarrow \rangle$ *orthogonal* iff there is some $A \in F$ such that $P(A) = 0$ but $P'(A) = 1$.

A model $\langle W, F, P, \rightarrow \rangle$ is *full* iff for all $A \in F$ and $r \in (0, P(A))$, there is some $B \subset A$ such that $P(B) = r$.

The second and third results make use of one or both of the following two constraints that might govern any given algebra $\langle W, F, \rightarrow \rangle$ or model $\langle W, F, P, \rightarrow \rangle$:

(L1) For all $A, B, [A \cap (A \rightarrow B)] \subseteq (AB)$.
(L2) For all $A, B, C, [(A \rightarrow B) \cap (A \rightarrow C)] \subseteq A \rightarrow (B \cap C)$.

I have chosen, for convenience, to treat probability functions as attaching values to propositions, as opposed to sentences that express propositions. However, we shall need sentences and interpretations thereof in order to give a clear statement of the third result. Henceforth, boldface capitals ($\mathbf{A}, \mathbf{B}, \mathbf{C}, \ldots$) will be used to denote atomic sentences, and boldface Greek capitals ($\mathbf{\Phi}, \mathbf{\Psi}, \ldots$) will be used to denote sentences generally. For the purposes of this chapter, a *sentence* will be understood to be a string of symbols taken from a language consisting of countably many atomic sentences, three truth-functional connectives ("&," "$\neg$", and "$\vee$"), parentheses, and one extra two-place connective "$\rightarrow$". Those strings of symbols that are to count as sentences are given by the following inductive definition:

All atomic sentences are sentences.
If $\mathbf{\Phi}$ and $\mathbf{\Psi}$ are sentences, then so are $\neg \mathbf{\Phi}, (\mathbf{\Phi} \& \mathbf{\Psi}), (\mathbf{\Phi} \vee \mathbf{\Psi})$, and $(\mathbf{\Phi} \rightarrow \mathbf{\Psi})$.

An *interpretation* is a pair $\langle I, M \rangle$, where M is a model $\langle W, \mathsf{F}, P, \rightarrow \rangle$ and I is a function that assigns to each sentence Φ an element $I(\Phi)$ of F, subject to the following restrictions:

$$I(\neg \Phi) = W - I(\Phi)$$
$$I(\Phi \,\&\, \Psi) = I(\Phi) \cap I(\Psi)$$
$$I(\Phi \vee \Psi) = I(\Phi) \cup I(\Psi)$$
$$I(\Phi \rightarrow \Psi) = I(\Phi) \rightarrow I(\Psi)$$

Finally, the probability of a sentence Φ relative to an interpretation $\langle I, M \rangle$ is simply $P(I(\Phi))$, where $P \in M$. For the sake of simplicity and at the price of a bit of harmless ambiguity, I shall often write the probability of a sentence Φ relative to an interpretation simply as $P(\Phi)$.

This completes the preliminaries.

THREE NEW RESULTS

The first result I shall prove shows that there are severe limitations on how many nontrivial CCCP-functions can be defined on any given algebra. It is this:

First result. If P and P' are nontrivial CCCP-functions for the same algebra, and if $P \neq P'$, then P and P' are orthogonal.

The second and third results are stronger than the first in that they even more severely restrict the sorts of models for which CCCP can hold. On the other hand, they are weaker in the sense that each assumes more about the "$\rightarrow$" than that it is simply a binary operator.

Second result. Any nontrivial model that obeys both CCCP and (L1) is full.

Call an interpretation $\langle I, M \rangle$ *allowable* iff CCCP, (L1), and (L2) all hold for M, and I interprets the atomic sentences $\mathbf{A}, \mathbf{B}$, and $\mathbf{C}$ as disjoint propositions with positive probability. Then we have the following:

Third result. There is an effective procedure for specifying, for any positive integer n, a set of n sentences $\{\Phi_1, \ldots, \Phi_n\}$ such that the only atomic sentences appearing in $\Phi_1, \ldots, \Phi_n$ are $\mathbf{A}, \mathbf{B}$, and $\mathbf{C}$, and such that the following hold, for any allowable interpretation:

(i) For any i, $P(\Phi_i) = 1/n$,
(ii) $P(\Phi_i \,\&\, \Phi_j) = 0$ if $i \neq j$.

143

The proof will proceed from the assumption that P and P' are nontrivial, nonidentical CCCP-functions for some algebra $\langle W, \mathsf{F}, \rightarrow \rangle$. First, more terminology:[3]

> P *dominates* P' *over* A iff for all $X, P(AX) \geqslant P'(AX)$.
> P *strongly dominates* P' *over* A iff for all $X, P(AX) > P'(AX)$ when $P'(AX) \neq 0$.
> P *matches* P' *over* A iff for all $X, P(AX) = P'(AX)$.
> P and P' are *nearly orthogonal* iff for all $\varepsilon > 0$ there is some A such that $P(A) > 1 - \varepsilon$ and $P'(A) < \varepsilon$ [equivalently: iff for all $\varepsilon > 0$ there is some A such that $P'(A) > 1 - \varepsilon$ and $P(A) < \varepsilon$].

The proof is long, but straightforward. I first establish (Lemma 0) a claim that is used at separate points later in the proof. I then show (Lemmas 1–2) that there are disjoint propositions A and B such that $P(A) > P'(A)$ and $P(B) > P'(B)$, and likewise that there are disjoint propositions A' and B' such that $P'(A') > P(A')$ and $P'(B') > P(B')$. It follows (Lemma 3) that P and P' are nearly orthogonal, and therefore (Lemma 4) that they are orthogonal.

Lemma 0. Suppose, for some disjoint $A, B \in \mathsf{F}$, that $P(A) > P'(A)$ and $P(B) > P'(B)$. Then for any $n > 0$ there is an X such that $P(X) > nP'(X)$. Likewise with P and P' reversed.

Proof. We prove only the first claim; the proof of the corresponding claim with P and P' reversed is exactly parallel. Suppose, then, that we are given disjoint $A, B \in \mathsf{F}$ such that $P(A) > P'(A)$ and $P(B) > P'(B)$.

If $P'(A) = 0$ or $P'(B) = 0$, the lemma follows immediately; so suppose otherwise. Because $P(A) > P'(A), P(\bar{A}) < P'(\bar{A})$; likewise, $P(\bar{B}) < P'(\bar{B})$. Further, neither $P(\bar{A})$ nor $P(\bar{B})$ is equal to zero, because A and B are disjoint and both have positive P-probability. So let $\delta = \min[P'(\bar{A})/P(\bar{A}), P'(\bar{B})/P(\bar{B})]$. Then δ is well defined, and $\delta > 1$. We now use induction to show that for each integer $i \geqslant 0$, there is a C_i such that (i) either $C_i \subseteq \bar{A}$ or $C_i \subseteq \bar{B}$, and (ii) $P(C_i) > \delta^i P'(C_i)$. Because $\delta > 1$, the lemma follows immediately.

Base step ($i = 0$): Let $C_0 = A$. Then (i) and (ii) are obviously satisfied.

Induction step ($i = n$): Suppose C_n meets conditions (i) and (ii). If $P'(C_n) = 0$, we are done; so suppose otherwise. Define the proposition D as follows: If $C_n \subseteq \bar{A}$, let $D = \bar{A} \rightarrow C_n$; otherwise, let $D = \bar{B} \rightarrow C_n$ (in which case it must be that $C_n \subseteq \bar{B}$). If $D = \bar{A} \rightarrow C_n$, then we have the following, using CCCP:

$$\frac{P(D)}{P'(D)} = \frac{P(C_n|\bar{A})}{P'(C_n|\bar{A})} = \frac{P(C_n)P'(\bar{A})}{P'(C_n)P(\bar{A})} > \frac{\delta^n P'(\bar{A})}{P(\bar{A})} \geqslant \delta^{n+1}$$

If $D = \bar{B} \to C_n$, then we have the following, again using CCCP:

$$\frac{P(D)}{P'(D)} = \frac{P(C_n|\bar{B})}{P'(C_n|\bar{B})} = \frac{P(C_n)P'(\bar{B})}{P'(C_n)P(\bar{B})} > \frac{\delta^n P'(\bar{B})}{P(\bar{B})} \geqslant \delta^{n+1}$$

It follows that $P(D) > \delta^{n+1} P'(D)$.

It therefore cannot be both that $P(D\bar{A}) \leqslant \delta^{n+1} P'(D\bar{A})$ and that $P(DA) \leqslant \delta^{n+1} P'(DA)$; else the foregoing inequality would not hold. So either $P(D\bar{A}) > \delta^{n+1} P'(D\bar{A})$ or $P(DA) > \delta^{n+1} P'(DA)$. If the first of these holds, let $C_{n+1} = D\bar{A}$; else let $C_{n+1} = DA$. Either way, $P(C_{n+1}) > \delta^{n+1} P'(C_{n+1})$, and either $C_{n+1} \subseteq \bar{A}$ or $C_{n+1} \subseteq \bar{B}$. Hence C_{n+1} meets conditions (i) and (ii). Q.E.D.

Lemma 1. For any A, if P dominates P' over A, then P strongly dominates P' over A. Likewise for P' dominating P.

Proof. We first show that for no A such that $P(A) > 0$ does P match P' over A. To do this, we show that for all A and X such that $P(A) \neq 0$, $P(X) = \sum_{i=0}^{\infty} P(\bar{A})^i P(AX_i)$, where X_i is specified via the following recursive definition: $X_0 = X$; $X_{i+1} = \bar{A} \to X_i$. It will be clear from the demonstration that the same equation holds with P replaced by P', provided again that $P'(A) \neq 0$.

Note first that CCCP entails that for all A and B, $P(AB) = P(A)P(A \to B)$. Likewise for P'. So let us first write $P(X) = P(AX) + P(\bar{A}X) = P(AX) + P(\bar{A})P(\bar{A} \to X)$. This equation can be rewritten thus:

$$P(X) = \sum_{i=0}^{0} P(\bar{A})^i P(AX_i) + P(\bar{A})P(X_1)$$

Now, suppose, for induction, that we have established

$$P(X) = \sum_{i=0}^{n} P(\bar{A})^i P(AX_i) + P(\bar{A})^{n+1} P(X_{n+1})$$

Then

$$P(X_{n+1}) = P(AX_{n+1}) + P(\bar{A}X_{n+1}) = P(AX_{n+1}) + P(\bar{A})P(\bar{A} \to X_{n+1})$$

whence

$$P(X) = \sum_{i=0}^{n+1} P(\bar{A})^i P(AX_i) + P(\bar{A})^{n+2} P(X_{n+2}).$$

145

Hence, by induction, the equation holds for all n. Finally, because $\lim_{n \to \infty} P(\bar{A})^{n+1} P(X_{n+1}) = 0$ [given that $P(\bar{A}) < 1$], we have

$$P(X) = \sum_{i=0}^{\infty} P(\bar{A})^i P(AX_i).$$

Clearly, the derivation holds for P' as well.

Now, suppose that there is some A such that P matches P' over A, and $P(A) > 0$. If $P(A) = 1$, then $P = P'$; thus $P(A) < 1$. Then for any X,

$$P(X) = \sum_{i=0}^{\infty} P(\bar{A})^i P(AX_i).$$

Because P matches P' over A, $P(A) = P'(A)$, whence $P(\bar{A}) = P'(\bar{A})$. Likewise, for all i, $P(AX_i) = P'(AX_i)$. It follows that

$$P(X) = \sum_{i=0}^{\infty} P'(\bar{A})^i P'(AX_i) = P'(X).$$

Because X was arbitrary, $P = P'$, contradicting the given.

Finally, suppose that for some A, P dominates P' over A, but P does not strongly dominate P' over A. Then there is an X such that $P(AX) = P'(AX) \neq 0$. Consider such an X, and let $B = AX$. Suppose that for some $Y, P(BY) > P'(BY)$. Because $P(B) = P'(B)$, it must be that $P(B\bar{Y}) < P'(B\bar{Y})$ – which contradicts the supposition that P dominates P' over A. Hence, for all Y, $P(BY) = P'(BY)$, whence P matches P' over B, which is impossible. Clearly the proof is symmetric in P and P'. Q.E.D.

Lemma 2. There are disjoint $A, B \in F$ such that $P(A) > P'(A)$, and $P(B) > P'(B)$. Likewise, there are disjoint $A', B' \in F$ such that $P'(A') > P(A')$, and $P'(B') > P(B')$.

Proof. We shall prove only the first claim; the second follows by exactly parallel reasoning. We proceed by *reductio* from the assumption that there are *not* disjoint $A, B \in F$ such that $P(A) > P'(A)$, and $P(B) > P'(B)$.

Consider the set of real numbers $S = \{r | \text{for some } X \text{ such that } P'$ dominates P over $X, r = P'(X)\}$; S is the set of P'-probabilities of those propositions over which P' dominates P. Let s be the least upper bound of S. If $s = 0$, then there must be disjoint A and B such that $P(A) > P'(A)$ and $P(B) > P'(B)$; thus $s > 0$. Because s is the least upper bound of S, there is a sequence of propositions $B_1, B_2, B_3, \ldots$ such that (i) for each i, P'

dominates P over B_i, and (ii) $\lim_{i \to \infty} P'(B_i) = s$. Let $C = \cup \{B_i\}$. $P'(C) \geqslant s$, and P' dominates P over C,[4] so in fact $P'(C) = s$. Because s is the least upper bound of S, for any $Y \subseteq \bar{C}, P'$ dominates P over Y iff $P'(Y) = P(Y) = 0$. And, from Lemma 1, P' strongly dominates P over C, hence $P'(C) > P(C)$, hence $P(\bar{C}) > P'(\bar{C})$. Further, $P'(\bar{C}) > 0$.[5]

We show now that $\bar{C}$ is both a P-atom and a P'-atom. Suppose first that $\bar{C}$ is not a P-atom. Then for some X, both $P(X\bar{C}) > 0$ and $P(\bar{X}\bar{C}) > 0$. Choose such an X. Then either $P(X\bar{C}) > P'(X\bar{C})$ or $P(\bar{X}\bar{C}) > P'(\bar{X}\bar{C})$. By hypothesis it cannot be both; so suppose instead (without loss of generality) that $P(X\bar{C}) > P'(X\bar{C})$ but $P(\bar{X}\bar{C}) \leqslant P'(\bar{X}\bar{C})$. Now, P' cannot dominate P over $\bar{X}\bar{C}$, because $P(\bar{X}\bar{C}) \neq 0$. So there must be some $Y \subset \bar{X}\bar{C}$ such that $P(Y) > P'(Y)$. But this contradicts our hypothesis. Finally, $\bar{C}$ must also be a P'-atom. For suppose not: Then for some $X, P'(X\bar{C}) > 0$ and $P'(\bar{X}\bar{C}) > 0$. But either $P(X\bar{C}) = 0$ or $P(\bar{X}\bar{C}) = 0$, because $\bar{C}$ is a P-atom. Assume without loss of generality that $P(X\bar{C}) = 0$. Then P' dominates P over $X\bar{C}$ and $P'(X\bar{C}) \neq 0$, which is impossible. So $\bar{C}$ is a P'-atom as well. In fact, we can demonstrate that the following hold, for any X:

(i) $P(X\bar{C}) = 0$ iff $P'(X\bar{C}) = 0$,
(ii) If $P(X) > P'(X)$, then $P(X\bar{C}) = P(\bar{C})$,
(iii) If $P(X) < P'(X)$, then $P(X\bar{C}) = 0$.
(iv) If $P(X) = P'(X)$, then either $P(X) = 0$ or $P(X) = 1$.

To establish (i), suppose first that $P(X\bar{C}) = 0$ but $P'(X\bar{C}) \neq 0$. Then P' dominates P over $X\bar{C}$ but $P'(X\bar{C}) > 0$, which is impossible. On the other hand, if $P(X\bar{C}) \neq 0$ but $P'(X\bar{C}) = 0$, then $P(\bar{X}\bar{C}) = 0$ but $P'(\bar{X}\bar{C}) \neq 0$, which is impossible. To establish (ii), suppose that $P(X) > P'(X)$ but $P(X\bar{C}) = 0$. Then $P'(X\bar{C}) = 0$, by (i). Then $P(XC) = P(X) > P'(X) = P'(XC)$, which is impossible, because P' dominates P over C. To establish (iii), suppose that $P(X) < P'(X)$ but $P(X\bar{C}) \neq 0$. Then $P(\bar{X}) > P'(\bar{X})$ and $P(\bar{X}\bar{C}) = 0$, which is impossible, by (ii). To establish (iv), suppose that $P(X) = P'(X)$ but $0 < P(X) < 1$. There are two cases. If $P(XC) = P(X)$, then $P(X\bar{C}) = 0 = P'(X\bar{C})$, hence $P(XC) = P(X) = P'(X) = P'(XC) > 0$, which is impossible. On the other hand, if $P(XC) < P(X)$, then $P(X\bar{C}) = P(\bar{C})$, whence $P(\bar{X}\bar{C}) = 0 = P'(\bar{X}\bar{C})$. But in that case, $P(\bar{X}C) = P(\bar{X}) = P'(\bar{X}) = P'(\bar{X}C) > 0$, which is impossible. Thus (i)–(iv) are established.

C cannot be a P'-atom, else P' is trivial. So there is a D such that $P'(CD) > 0$ and $P'(C\bar{D}) > 0$. Further, because P' strongly dominates P over $C, P'(CD) > P(CD)$ and $P'(C\bar{D}) > P(C\bar{D})$. Then, by Lemma 0, for any $n > 0$ there is an X such that $P'(X) > nP(X)$. Now consider some arbitrary $n > 1$, and consider the set of real numbers $T = \{r | \text{for some } X, r = P'(X) > nP(X)\}$. Let t be the least upper bound of T. Clearly $t > 0$; we now show that in fact $t \geqslant P'(C)$.

147

By the definition of t, for any $\varepsilon > 0$ there is an X such that (i) $t \geqslant P'(X) > t - \varepsilon$, and (ii) $P'(X) > nP(X)$. Suppose that $t < P'(C)$. Then choose $\varepsilon < t(P'(C) - t)$, and consider some X that satisfies (i) and (ii) for such a choice of ε. Then $P'(X\bar{C}) = 0 = P(X\bar{C})$, by (i) and (iii). By (iii) and (iv), $P(X \cup \bar{C}) > P'(X \cup \bar{C})$. So let $E = (X \cup \bar{C}) \to X$, and suppose that $P'(E) \leqslant nP(E)$. Employing CCCP and rearranging, this yields $P'(X)P(X \cup \bar{C}) \leqslant nP(X)P'(X \cup \bar{C})$, which is impossible. So $P'(E) > nP(E)$; therefore $P'(E) \leqslant t$. Thus, by CCCP, $P'(X) \leqslant tP'(X \cup \bar{C}) = t(P'(X) + P'(\bar{C})) \leqslant t(t + P'(\bar{C}))$. But $t - \varepsilon < P'(X)$, whence $t - \varepsilon < t(t + P'(\bar{C}))$, whence $\varepsilon > t - t(t + P'(\bar{C})) = t(P'(C) - t)$. And this contradicts our choice of ε. By *reductio*, then, $t \geqslant P'(C)$.

But this is impossible, for the choice of n was arbitrary. So let us make n quite large – specifically, let it be the case that $1/n < P(C)$ [possible, because nontriviality requires that $P(C) > 0$]. Because $t \geqslant P'(C)$, for any $\varepsilon > 0$ there is an X such that $P'(X) > nP(X)$ and $P'(X) > P'(C) - \varepsilon$. Again, for any such X, $P'(X\bar{C}) = 0 = P(X\bar{C})$, whence $P'(XC) = P'(X)$. So $P'(\bar{X}C) < \varepsilon$. Because $P'(\bar{X}C) \geqslant P(\bar{X}C)$, $P(\bar{X}C) < \varepsilon$. Because $P'(X) > nP(X)$, $P(X) < 1/n$. But $P(XC) = P(X)$. Therefore $P(C) = P(XC) + P(\bar{X}C) = P(X) + P(\bar{X}C) < (1/n) + \varepsilon$. But this is so for arbitrary $\varepsilon > 0$; therefore, in fact, $P(C) \leqslant 1/n$ – a contradiction. This completes the *reductio*. Q.E.D.

Lemma 3. If there are disjoint A, B such that $P(A) > P'(A)$, and $P(B) > P'(B)$, and disjoint A', B' such that $P'(A') > P(A')$, and $P'(B') > P(B')$, then P and P' are nearly orthogonal.

Proof. Suppose that there are the needed disjoint propositions. Then given arbitrary $\varepsilon > 0$, we must find a C such that $P(C) > 1 - \varepsilon$ and $P'(C) < \varepsilon$.

Lemma 0 guarantees that for any $n > 1$, there are X and Y such that $P(X) > nP'(X)$ and $P'(Y) > nP(Y)$. Then suppose we are given $\varepsilon > 0$; choose n such that $2/n < \varepsilon$. Consider the sets of real numbers $S = \{r | \text{for some } X, r = P(X) > nP'(X)\}$ and $S' = \{r | \text{for some } Y, r = P'(Y) > nP(Y)\}$. Let s and s' be the least upper bounds of S and S', respectively; clearly $s \neq 0 \neq s'$. We first show that in fact either $s > 1 - (2/n)$ or $s' > 1 - (2/n)$.

Choose some $\delta > 0$ such that $\delta < s/n$ and $\delta < s'/n$. By the definitions of s and s', there are X and Y such that $s - \delta < P(X) \leqslant s, s' - \delta < P'(Y) \leqslant s'$, $P(X) > nP'(X)$, and $P'(Y) > nP(Y)$. Choose such. Suppose now that $P((X \cup Y) \to X) \leqslant nP'((X \cup Y) \to X)$ and $P'((X \cup Y) \to Y) \leqslant nP((X \cup Y) \to Y)$. Then all terms must be positive, so we can multiply these inequalities to yield

$$P((X \cup Y) \to X)P'((X \cup Y) \to Y) \leqslant n^2 P'((X \cup Y) \to X)P((X \cup Y) \to Y)$$

148

Using CCCP, this becomes

$$\left(\frac{P(X)}{P(X\cup Y)}\right)\left(\frac{P'(Y)}{P'(X\cup Y)}\right)\leqslant n^2\left(\frac{P'(X)}{P'(X\cup Y)}\right)\left(\frac{P(Y)}{P(X\cup Y)}\right)$$

Simplifying, we deduce that $P(X)P'(Y)\leqslant n^2 P'(X)P(Y)$ – which contradicts the supposition that $P(X)>nP'(X)$ and $P'(Y)>nP(Y)$. So either

(i) $P((X\cup Y)\to X)>nP'((X\cup Y)\to X)$, or
(ii) $P'((X\cup Y)\to Y)>nP((X\cup Y)\to Y)$.

Suppose (i) holds. Then $P((X\cup Y)\to X)\leqslant s$, whence, by CCCP, $P(X)\leqslant sP(X\cup Y)\leqslant s(P(X)+P(Y))$. But $nP(Y)<P'(Y)\leqslant 1$ and $P(X)\leqslant s$; therefore $s-\delta<P(X)<s(s+(1/n))$. By the choice of δ, it follows that $s-(s/n)<s(s+1/n))$; therefore $s>1-(2/n)$. Likewise, if (ii) holds, then $s'>1-(2/n)$.

So either $s>1-(2/n)>1-\varepsilon$ or $s'>1-(2/n)>1-\varepsilon$. If the first holds, there is an X such that $1-\varepsilon<P(X)\leqslant s$ and $P(X)>nP'(X)$, whence $P'(X)<1/n<\varepsilon$. This X is the needed proposition C. If the second holds, then there is a Y such that $1-\varepsilon<P'(Y)\leqslant s$ and $P'(Y)>nP(Y)$, whence $P(Y)<1/n<\varepsilon$. But in this case, $P(\bar{Y})>1-\varepsilon$ and $P'(\bar{Y})<\varepsilon$. Thus, $\bar{Y}$ is the needed proposition C. One way or the other, C exists. Q.E.D.

Lemma 4. *If P and P' are nearly orthogonal, then they are orthogonal.*

Proof. Suppose that P and P' are nearly orthogonal. We use induction to show that for any $\varepsilon>0$, there is a sequence of propositions $A_0, A_1,\dots$ such that the following hold, for each i: (i) if $j<i$, then $A_i\subseteq A_j$; (ii) $P(A_i)>1-2\varepsilon+(\varepsilon/2^i)$; (iii) $P'(A_i)<\varepsilon/2^i$. So suppose we are given some $\varepsilon>0$; then, by the near orthogonality of P and P', there is an A such that $P(A)>1-\varepsilon$ and $P'(A)<\varepsilon$. Choosing such an A, we proceed with the induction:

Base step: Let $A_0=A$. Then (i)–(iii) are obviously satisfied.

Induction step: Suppose A_n has been defined and satisfies (i)–(iii). Since P and P' are nearly orthogonal, there is a B such that $P(B)>1-\varepsilon/2^{n+1}$ and $P'(B)<\varepsilon/2^{n+1}$. Choose such a B, and let $A_{n+1}=A_nB$. Then clearly (i) and (iii) hold. Further, $P(A_nB)\geqslant P(A_n)-P(\bar{B})>1-2\varepsilon+\varepsilon/2^n-\varepsilon/2^{n+1}=1-2\varepsilon+\varepsilon/2^{n+1}$. So (ii) holds as well. This completes the induction.

Let $C=\cap\{A_i\}$. Then $P(C)=\lim_{i\to\infty}P(A_i)\geqslant 1-2\varepsilon$ and $P'(C)=\lim_{i\to\infty}P'(A_i)=0$. So for any $\varepsilon>0$ there is a C such that $P(C)\geqslant 1-2\varepsilon$ and $P'(C)=0$. So in fact there is a sequence of propositions $C_1, C_2,\dots$ such that for each i, $P(C_i)\geqslant 1-(1/i)$ and $P'(C_i)=0$. Let $D=\cup\{C_i\}$. Then $P(D)=1$ but $P'(D)=0$, whence P and P' are orthogonal. Q.E.D.

From Lemmas 2, 3, and 4, it immediately follows that P and P' are orthogonal. Q.E.D.

The proof consists in showing that no nontrivial model $\langle W, \mathsf{F}, P, \rightarrow \rangle$ that obeys CCCP and (L1) can contain a P-atom. Because a model is atomless iff it is full,[6] the second result follows. The proof proceeds by *reductio*:

Suppose that CCCP and (L1) hold for a nontrivial model $\langle W, \mathsf{F}, P, \rightarrow \rangle$ and that $\bar{A} \in \mathsf{F}$ is a P-atom. Then A is not a P-atom, else the model is trivial. So there is a B such that $0 < P(AB) < P(A)$. Choose such a B. Then we have the following.

$$P(A \rightarrow B) = P(A \cap A \rightarrow B) + P(\bar{A} \cap A \rightarrow B)$$
$$\leqslant P(AB) + P(\bar{A} \cap A \rightarrow B) \qquad \text{[by (L1)]}$$

Using CCCP and rearranging, it follows that $P(B|A)P(\bar{A}) \leqslant P(\bar{A} \cap A \rightarrow B)$.

Because $\bar{A}$ is a P-atom, the right-hand side is equal to either 0 or $P(\bar{A})$. But it cannot be the former, because the left-hand side is positive. Thus, we have established

(i) $P(\bar{A} \cap A \rightarrow B) = P(\bar{A})$.

An exactly parallel argument yields [given that $P(A\bar{B}) > 0$]

(ii) $P(\bar{A} \cap A \rightarrow \bar{B}) = P(\bar{A})$.

Finally, (i) and (ii) together yield

(iii) $P(\bar{A} \cap A \rightarrow B \cap A \rightarrow \bar{B}) = P(\bar{A})$.[7]

Note now that

$$P(A \rightarrow B \cap A \rightarrow \bar{B}) = P(A \cap A \rightarrow B \cap A \rightarrow \bar{B}) + P(\bar{A} \cap A \rightarrow B \cap A \rightarrow \bar{B})$$

The first term on the right-hand side vanishes, thanks to (L1). Employing (iii), we therefore have $P(A \rightarrow B \cap A \rightarrow \bar{B}) = P(\bar{A})$. Because $P(A \rightarrow B) \geqslant P(A \rightarrow B \cap A \rightarrow \bar{B})$, it follows, using CCCP and a rearrangement, that

(iv) $P(AB) \geqslant P(A)P(\bar{A})$.

Now, (iv) holds for *any* B such that $0 < P(AB) < P(A)$. So A must contain a P-atom. For suppose not: Then choose some $C_0 \subset A$; because C_0 is not a P-atom, there is a D such that $0 < P(C_0 D)$ and $0 < P(C_0 \bar{D})$. But either $P(C_0 D) \leqslant \frac{1}{2}P(C_0)$ or $P(C_0 \bar{D}) \leqslant \frac{1}{2}P(C_0)$. So – on the assumption that A contains no P-atoms – we can choose a $C_1 \subset C_0$ such that $0 < P(C_1) \leqslant \frac{1}{2}P(C_0)$, and likewise – because this C_1 contains no P-atoms – a $C_2 \subset C_1$

such that $0 < P(C_2) \leqslant \frac{1}{2}P(C_1)$, and so on. In other words, A will contain an infinite sequence of propositions $C_0, C_1, C_2, \ldots$ such that $0 < P(C_i) \leqslant (1/2^i)P(C_0)$. But (iv) requires that for all such C_i, $P(C_i) \geqslant P(A)P(\bar{A})$, which is impossible.

Thus, let $C \subset A$ be a P-atom. Then, because $A \cap A \rightarrow C \subseteq C$, $P(A \cap A \rightarrow C) = 0$ or $P(A \cap A \rightarrow C) = P(C)$. If the latter, then $P(C|A) = P(A \rightarrow C) = P(\bar{A}) + P(C)$, which by a rearrangement yields $P(C) = P(A)$ – a contradiction. So $P(A \cap A \rightarrow C) = 0$, whence $P(C|A) = P(\bar{A})$, whence $P(C) = P(A)P(\bar{A})$. So any P-atom within A has probability $P(A)P(\bar{A})$.

Now, the choice of A was *arbitrary*, subject only to the restriction that $\bar{A}$ was a P-atom. Hence, the argument so far establishes that for *any* $X \in F$, if $\bar{X}$ is a P-atom, then any P-atom contained in X has probability $P(X)P(\bar{X})$. Hence in particular – letting $C \subseteq A$ be a P-atom – any atom contained in $\bar{C}$ has probability $P(C)P(\bar{C}) = P(A)P(\bar{A})[1 - P(A)P(\bar{A})]$.

But $\bar{A}$ is one such P-atom. Thus $P(\bar{A}) = P(A)P(\bar{A})[1 - P(A)P(\bar{A})]$, which is clearly impossible. This completes the *reductio*. Q.E.D.

PROOF OF THIRD RESULT

Recall that an interpretation $\langle I, M \rangle$ is *allowable* iff I assigns to the atomic sentences $\mathbf{A}, \mathbf{B}$, and $\mathbf{C}$ disjoint propositions with positive probability, and CCCP, (L1), and (L2) all hold for M. We need to detail a method that, given any positive integer n, will yield a set of sentences $\{\Phi_1, \ldots, \Phi_n\}$ whose sole atomic constituents are $\mathbf{A}, \mathbf{B}$, and $\mathbf{C}$ and such that, relative to any allowable interpretation, $P(\Phi_i) = 1/n$ and $P(\Phi_i \& \Phi_j) = 0$ when $i \neq j$. The case $n = 1$ is trivial; for the cases $n = 2$ and $n = 3$, I shall give the needed sentences explicitly. The remainder of the proof will be an induction, using the case $n = 3$ as the base step.

In Section 1 of the Appendix, it is demonstrated that given a model $\langle W, F, P, \rightarrow \rangle$ for which CCCP, (L1), and (L2) all hold, $P(A \cap A \rightarrow B) = P(AB)$ for any A and B. Given CCCP, it follows at once that $P(A \rightarrow B|A) = P(A \rightarrow B)$ when $P(A) > 0$, hence that $P(A \rightarrow B|\bar{A}) = P(A \rightarrow B)$, hence that

$$(*) P(\bar{A} \cap A \rightarrow B) = P(\bar{A})P(A \rightarrow B)$$

We shall make frequent use of $(*)$ in the proof that follows. Note that it holds even when $P(A) = 0$.

Let $\langle I, M \rangle$ be an allowable interpretation. Define the sentences Γ_0 and Γ_1 thus:

$$\Gamma_0 = (\mathbf{C} \& (\neg \mathbf{C} \rightarrow (\mathbf{B} \& (\neg \mathbf{B} \rightarrow \mathbf{A}))))$$
$$\Gamma_1 = (\mathbf{B} \& (\neg \mathbf{B} \rightarrow (\mathbf{C} \& (\neg \mathbf{C} \rightarrow \mathbf{A}))))$$

151

Let $A = I(\mathbf{A})$, $B = I(\mathbf{B})$, and $C = I(\mathbf{C})$. Then A, B, and C are pairwise disjoint. Thus,

$$
\begin{aligned}
P(\Gamma_0) &= P(C \cap (\bar{C} \to (B \cap (\bar{B} \to A)))) \\
&= P(C)P(\bar{C} \to (B \cap (\bar{B} \to A))) &&\text{[by (*)]} \\
&= P(C)P(B \cap (\bar{B} \to A) | \bar{C}) &&\text{(by CCCP)} \\
&= \frac{P(C)}{P(\bar{C})}P(B \cap (\bar{B} \to A)) &&\text{(because } B \subset \bar{C}) \\
&= \frac{P(C)P(B)}{P(\bar{C})}P(\bar{B} \to A) &&\text{[by (*)]} \\
&= \frac{P(C)P(B)}{P(\bar{C})}P(A | \bar{B}) &&\text{(by CCCP)} \\
&= \frac{P(A)P(B)P(C)}{P(\bar{B})P(\bar{C})} &&\text{(because } A \subset \bar{B})
\end{aligned}
$$

Similarly, it is easy to verify that

$$
P(\Gamma_1) = \frac{P(A)P(B)P(C)}{P(\bar{B})P(\bar{C})}
$$

Hence, $P(\Gamma_0) = P(\Gamma_1)$. Let $\Phi_0 = ((\Gamma_0 \vee \Gamma_1) \to \Gamma_0)$, $\Phi_1 = ((\Gamma_0 \vee \Gamma_1) \to \Gamma_1)$. Let $G_0 = I(G_0)$, $G_1 = I(\Gamma_1)$; thus $P(G_0) = P(G_1) > 0$, and $G_0 G_1 = \varnothing$. We show that $P(\Phi_0) = P(\Phi_1) = \frac{1}{2}$:

$$
\begin{aligned}
P(\Phi_0) &= P((G_0 \cup G_1) \to G_0) \\
&= P(G_0 | G_0 \cup G_1) &&\text{(by CCCP)} \\
&= \frac{P(G_0)}{P(G_0 \cup G_1)} &&\text{[because } G_0 \subset (G_0 \cup G_1)] \\
&= \frac{P(G_0)}{P(G_0) + P(G_1)} &&\text{(because } G_0 G_1 = \varnothing) \\
&= \tfrac{1}{2} &&\text{[because } P(G_0) = P(G_1)]
\end{aligned}
$$

It is easy to verify, in just the same way, that $P(\Phi_1) = \frac{1}{2}$. Further, $P(\Phi_0 \& \Phi_1) = 0$:

$$
\begin{aligned}
P(\Phi_0 \& \Phi_1) &= P(((G_0 \cup G_1) \to G_0) \cap ((G_0 \cup G_1) \to G_1)) \\
&\leqslant P((G_0 \cup G_1) \to (G_0 G_1)) &&\text{[by (L1)]} \\
&= P(G_0 G_1 | G_0 \cup G_1) &&\text{(by CCCP)} \\
&= 0 &&\text{(because } G_0 G_1 = \varnothing)
\end{aligned}
$$

This handles the case $n = 2$; we turn now to the case $n = 3$.

Notice that while $G_0 \subseteq C, 0 < P(G_0) < P(C)$; hence $P(C\bar{G}_0) > 0$. Likewise, $P(B\bar{G}_1) > 0$. So let $\Gamma_2 = (\neg\Gamma_0 \& C)$; let $\Gamma_3 = (\neg\Gamma_1 \& B)$. Let $G_2 = I(\Gamma_2)$ and $G_3 = I(\Gamma_3)$. Then the G_i are four pairwise disjoint propositions with positive probability. We define three more sentences in terms of the Γ_i:

$$X_0 = (\Gamma_0 \& (\neg\Gamma_0 \to (\Gamma_1 \& (\neg\Gamma_1 \to (\Gamma_2 \& (\neg\Gamma_2 \to \Gamma_3))))))$$
$$X_1 = (\Gamma_1 \& (\neg\Gamma_1 \to (\Gamma_2 \& (\neg\Gamma_2 \to (\Gamma_0 \& (\neg\Gamma_0 \to \Gamma_3))))))$$
$$X_2 = (\Gamma_2 \& (\neg\Gamma_2 \to (\Gamma_0 \& (\neg\Gamma_0 \to (\Gamma_1 \& (\neg\Gamma_1 \to \Gamma_3))))))$$

It is now routine to verify, using CCCP and (∗), that

$$P(X_0) = P(X_1) = P(X_2) = \frac{P(G_0)P(G_1)P(G_2)P(G_3)}{P(\bar{G}_0)P(\bar{G}_1)P(\bar{G}_2)}$$

We now define[8]

$$\Psi_0 = ((X_0 \vee X_1 \vee X_2) \to X_0)$$
$$\Psi_1 = ((X_0 \vee X_1 \vee X_2) \to X_1)$$
$$\Psi_2 = ((X_0 \vee X_1 \vee X_2) \to X_2)$$

Noting that $I(X_i \& X_j) = \varnothing$ when $i \neq j$, it follows, in a manner exactly analogous to the demonstration for the case $n = 2$, that $P(\Psi_0) = P(\Psi_1) = P(\Psi_2) = \frac{1}{3}$ and that $P(\Psi_i \& \Psi_j) = 0$ when $i \neq j$. This handles the case $n = 3$.

We now proceed with the induction. Suppose that, for $n \geqslant 3$, the set of sentences $\{\Phi_1, \ldots, \Phi_n\}$ satisfies the given conditions: (i) each Φ_i has just A, B, and C as its sole atomic constituents; (ii) for each $i, P(\Phi_i) = 1/n$; (iii) $P(\Phi_i \& \Phi_j) = 0$ when $i \neq j$. Then we need to show how to construct a set of sentences $\{\Psi_1, \ldots, \Psi_{n+1}\}$ that will satisfy these same conditions (with n replaced by $n + 1$). Let us begin with the following definitions:

Let Δ_k be a disjunction of all the Φ_i *except* Φ_k. There will be n such sentences (ignoring disjunctions that differ only in the order of the disjuncts). Note that, relative to any allowable interpretation, $P(\Phi_k) = P(\neg\Delta_k) = P(\Phi_k \& \neg\Delta_k)$. Let X_k^j be $(\neg\Delta_k \& (\Delta_k \to \Phi_j))$, where we require that $j \neq k$. There will be $n(n-1)$ such sentences.

Then, for all j, k such that $j \neq k, P(X_k^j) = P(\neg\Delta_k)P(\Delta_k \to \Phi_j)$; this follows by (∗). Thanks to the definitions and the properties of the $\Phi_i, P(\neg\Delta_k) = P(\Phi_k) = 1/n$ and $P(\Delta_k \to \Phi_j) = 1/(n-1)$. Hence, for all j, k such that $j \neq k$, $P(X_k^j) = 1/(n(n-1))$.

Further, $P(X_k^j \& X_m^l) = 0$ if $j \neq l$ or $k \neq m$. Suppose first that $k \neq m$. Then

$$P(X_k^j \& X_m^l) \leqslant P(\neg\Delta_k \& \neg\Delta_m)$$
$$= P(\neg\Delta_k) - P(\neg\Delta_k \& \Delta_m)$$
$$= P(\Phi_k) - P(\Phi_k)$$
$$= 0$$

153

Suppose next that $k = m$ but $j \neq l$. Then

$$
\begin{aligned}
P(X_k^j \& X_m^l) &\leqslant P((\Delta_k \to \Phi_j) \& (\Delta_k \to \Phi_l)) \\
&\leqslant P(\Delta_k \to (\Phi_j \& \Phi_l)) \\
&= P(\Phi_j \& \Phi_l | \Delta_k) \\
&= 0
\end{aligned}
$$

Thus, the set of sentences $\{X_k^j\}$ satisfies conditions (i)–(iii) for $n(n-1)$. Further, because $n \geqslant 3$, the number of sentences in this set must be greater than $n + 1$. So let the set $\{T_1, \ldots, T_{n+1}\}$ be a subset of the set $\{X_k^j\}$. Let the sentence Σ be a disjunction of all the T_i. Finally, let $\Psi_i = (\Sigma \to T_i)$. It is routine to apply the techniques used earlier to verify that the set $\{\Psi_1, \ldots, \Psi_{n+1}\}$ satisfies conditions (i)–(iii) for the case $n + 1$. Q.E.D.

DISCUSSION

Here I discuss corollaries, extensions, and limitations of my results. En route I also offer comparisons of the results in this chapter to some of the other work that has been done on CCCP.

First result

Given any algebra $\langle W, \mathsf{F}, \to \rangle$, we now know that any two nontrivial CCCP-functions for it are orthogonal. It follows automatically that the set of CCCP-functions for the algebra does not contain all probability functions definable on it, that the set is closed under conditioning only if it contains just trivial functions, that it is closed under conditioning restricted to a single partition of propositions only if it contains just trivial functions, and that it is closed under Jeffrey conditioning only if it contains just trivial functions. These four results – which are, of course, David Lewis's four triviality results – follow from the fact that neither conditioning nor Jeffrey conditioning ever maps a probability function onto an orthogonal probability function. Indeed, this will be so for *any* plausible "updating" rule. So in fact we have this corollary, much stronger than Lewis's results: If (nontrivial) $\langle W, \mathsf{F}, P, \to \rangle$ obeys CCCP, and P' is derived from P via some (reasonable) updating rule, then $\langle W, \mathsf{F}, P', \to \rangle$ does not obey CCCP.

We can state the first result as a kind of *uniqueness* claim: Given CCCP and an algebra $\langle W, \mathsf{F}, \to \rangle$, there is, at most, one way to choose (nontrivial) P, up to orthogonality. Further, we can dispense with the orthogonality qualification for a wide range of cases. Specifically, suppose that W is countably infinite, that to each world in W a non-zero "weight" is assigned, and that the probability of any proposition is simply the sum of the weights of those worlds that belong to it. Suppose further that there is some assignment of weights so that the resulting nontrivial model $\langle W, F, P, \rightarrow \rangle$ obeys CCCP. Then there is no function $P' \neq P$ such that P' is orthogonal to P, for that function would have to give probability 0 to some proposition to which P gives probability 1 – which is impossible, because the only such proposition is W itself. It follows, then, that P is the *unique* nontrivial function that satisfies CCCP for the given algebra.

On the other hand, we cannot in general dispense with the orthogonality qualification. For suppose that $< W_1, F_1, P_1, \rightarrow_1 >$ and $\langle W_2, F_2, P_2, \rightarrow_2 \rangle$ both obey CCCP. Without loss of generality, we can assume that W_1 and W_2 do not overlap. Then let $W^* = W_1 \cup W_2$. Let F^* be the closure under the Boolean operations (including countable union) of $F_1 \cup F_2$. Define the operation "$\rightarrow$*" on F^* thus: For any $A, B \in F^*$, $A \rightarrow^* B = (AW_1 \rightarrow_1 BW_1) \cup (AW_2 \rightarrow_2 BW_2)$. Finally, define two probability functions P_1^* and P_2^* thus: For all $A \in F^*$, $P_1^*(A) = P_1(AW_1)$ and $P_2^*(A) = P_2(AW_2)$. P_1^* and P_2^* are therefore both defined on the algebra $\langle W^*, F^*, \rightarrow^* \rangle$, and are orthogonal. Given any $A, B \in F^*$ such that $P_1^*(A) \neq 0$, we have

$$
\begin{aligned}
P_1^*(A \rightarrow^* B) &= P_1^*((AW_1 \rightarrow_1 BW_1) \cup (AW_2 \rightarrow_2 BW_2)) \\
&= P_1^*(AW_1 \rightarrow_1 BW_1) + P_1^*(AW_2 \rightarrow_2 BW_2) \\
&= P_1^*(AW_1 \rightarrow_1 BW_1) \\
&= P_1(AW_1 \rightarrow_1 BW_1) \\
&= P_1(BW_1 | AW_1) \\
&= P_1^*(B | A)
\end{aligned}
$$

Thus, CCCP holds for $\langle W^*, F^*, P_1^*, \rightarrow^* \rangle$. Likewise, it holds for $\langle W^*, F^*, P_2^*, \rightarrow^* \rangle$. In general, then, given any two models for which CCCP holds, there will be two other CCCP-models that employ the same arrow and whose probability functions are orthogonal. Indeed, it is easy to see that the method used for constructing these models can be extended indefinitely: Any countable set of nontrivial CCCP-models can be mapped one-to-one by a similar construction into a set of nontrivial, pairwise orthogonal CCCP-functions for a single algebra.

Second result

The second result is usefully compared to Hájek's 1989 result.[9] There he proves, without making any assumptions about the logical structure of the "$\rightarrow$," that any CCCP-function for any algebra must have an infinite range. Put another way, the result shows that no probability space $\langle W, F, P \rangle$ such that P has a finite range can be extended to a model for

155

which CCCP holds – even if we require *only* that the "→" be a binary operator. Note that this result cannot easily be strengthened, for if the range of $P = [0, 1]$, then of course there will be some "→" such that CCCP holds for $\langle W, \mathsf{F}, P, \rightarrow \rangle$; for each A and B, simply let $A \rightarrow B$ be some proposition whose probability is $P(B|A)$.[10] Suppose, however, that we wish to choose some "→" operation that behaves like a *conditional* – at least to the extent of obeying *modus ponens* (L1). Then the second result shows that the probability space $\langle W, \mathsf{F}, P \rangle$ with which we begin not only must be such that P has an infinite range but also must be full (from which it follows that W must be non-denumerable).

Third result

It follows at once from the third result that there is a method for producing, for any rational number $r \in [0, 1]$, a sentence Ψ that contains just **A, B**, and **C** as its atomic constituents, and such that $P(\Psi) = r$ relative to any allowable interpretation. For let $r = m/n$, where m and n are positive integers, and consider some intended interpretation $\langle I, M \rangle$ with $P \in M$. Again, by the third result, there is a set of n sentences $\{\Phi_i\}$ such that $P(\Phi_i) = 1/n$ and $P(\Phi_i \Phi_j) = 0$ when $i \neq j$. Let Ψ be a disjunction of m of these sentences. Then clearly $P(\Psi) = r$.

In addition, one of the restrictions on allowable interpretations can be dispensed with – although for the sake of readability I chose not to. Specifically, the proof of the third result is not affected if we replace the atomic sentences **A, B**, and **C** with nonconditional sentences whose syntax guarantees that they will be interpreted as disjoint [e.g., "(A & B)," "(A & ¬ B)," and "¬ A"]. Of course, there is no such syntactical way to guarantee, in addition, that all three will receive positive probability.

First and second results, combined

I noted earlier that the first result can be stated as a qualified uniqueness result. The qualification of orthogonality can be dropped in some more cases, provided we take into account the second result. Begin with the observation that if a model $\langle W, \mathsf{F}, P, \rightarrow \rangle$ is *regular* [i.e., $P(X) = 0$ iff $X = \varnothing$], then by the first result alone P is the unique CCCP-function for the algebra $\langle W, \mathsf{F}, \rightarrow \rangle$. Regularity is hard to come by if W is non-denumerable, but a weaker condition will suffice, once we assume (L1): Call a model $\langle W, \mathsf{F}, P, \rightarrow \rangle$ *quasi-regular* iff for all $X \in \mathsf{F}$, if $P(X) = 0$, then X contains, at most, countably many worlds. Thus, models can be quasi-regular even though W is non-denumerable. Suppose, then, that P' and P are distinct, nontrivial CCCP-functions for the same algebra

$\langle W, \mathsf{F}, \rightarrow \rangle$. Suppose too that $\langle W, \mathsf{F}, P, \rightarrow \rangle$ is quasi-regular. P and P' are orthogonal, whence P' must assign probability 1 to some proposition containing just countably many worlds. Then, because P' is countably additive and F is closed under countable unions, $\langle W, \mathsf{F}, P', \rightarrow \rangle$ must contain at least one atom. So $\langle W, \mathsf{F}, P', \rightarrow \rangle$ is not full, *contra* the second result. Hence, if the algebra $\langle W, \mathsf{F}, \rightarrow \rangle$ obeys (L1), and if P is a nontrivial CCCP-function for this algebra such that $\langle W, \mathsf{F}, P, \rightarrow \rangle$ is quasi-regular, then P is the unique nontrivial CCCP-function for this algebra.

APPENDIX

1. Proof that $P(A \cap A \rightarrow B) = P(AB)$

For convenience, I shall, in what follows, sometimes use "$\neg(—)$" to denote the set-theoretic complement of $(—)$. Here is the proof of the foregoing equation, given (L1), (L2), and CCCP.

Note first that (i) $P(A \rightarrow B \cup A \rightarrow \bar{B}) = 1$, as follows:

$$P(A \rightarrow B \cup A \rightarrow \bar{B}) = P(A \rightarrow B) + P(A \rightarrow \bar{B}) - P(A \rightarrow B \cap A \rightarrow \bar{B})$$
$$= P(B|A) + P(\bar{B}|A) - P(A \rightarrow B \cap A \rightarrow \bar{B}) \quad \text{(by CCCP)}$$
$$= 1 - P(A \rightarrow B \cap A \rightarrow \bar{B})$$
$$= 1 \qquad\qquad\qquad\qquad\qquad \text{[by (L2) and CCCP]}$$

Now,

$$P(A \cap A \rightarrow B) = P(A \cap A \rightarrow B \cap B) + P(A \cap A \rightarrow B \cap \bar{B})$$

But, by (L1), $(A \cap A \rightarrow B \cap \bar{B}) \subseteq (A \cap B \cap \bar{B}) = \varnothing$; thus

$$P(A \cap A \rightarrow B) = P(A \cap A \rightarrow B \cap B)$$
$$= P(AB) - P[AB \cap \neg(A \rightarrow B)]$$

Thus, it suffices to show that $P[AB \cap \neg(A \rightarrow B)] = 0$. Using (i), we have

$$P[AB \cap \neg(A \rightarrow B)] = P[AB \cap \neg(A \rightarrow B) \cap (A \rightarrow B \cup A \rightarrow \bar{B})]$$
$$= P[AB \cap \neg(A \rightarrow B) \cap A \rightarrow B]$$
$$+ P[AB \cap \neg(A \rightarrow B) \cap A \rightarrow \bar{B}]$$
$$= P[AB \cap \neg(A \rightarrow B) \cap A \rightarrow \bar{B}]$$
$$= 0 \quad \text{[by (L1)]}$$

Q.E.D.

2. Proof that full = atomless

We wish to show that any probability space $\langle W, \mathsf{F}, P \rangle$ is full iff it contains no atoms. Recall that a space is full iff for every $A \in \mathsf{F}$ and $r \in (0, P(A))$

157

there is a $B \subset A$ such that $P(B) = r$. So one direction is easy: If a space is full, then obviously it contains no atoms. For the other direction, we shall need the notion of a *gap*:

A *gap* (a, b) within a proposition A is an open interval of real numbers (a, b) such that $0 \leqslant a < b \leqslant P(A)$, and for no $r \in (a, b)$ is there a $B \subseteq A$ such that $P(B) = r$. We say that A *contains a gap* just in case there is some gap within A. To establish the proof that a space is full if atomless, we need three lemmas:

Lemma 1. If a proposition $A \in F$ contains no P-atoms, then A contains no gaps of the form $(0, b)$.

Proof. Suppose otherwise. We show by induction that there is an infinite sequence of propositions $A_0, A_1, \ldots$ such that the following hold, for each i: (i) $A_i \subseteq A$; (ii) $0 < P(A_i) \leqslant P(A)/2^i$.

Base step: Let $A_0 = A$. Then conditions (i) and (ii) are clearly satisfied.

Induction step: Suppose A_n exists and satisfies the two conditions. Because A_n is not a P-atom, there is some B such that both $0 < P(A_n B)$ and $0 < P(A_n \bar{B})$. Further, it cannot be that both $P(A_n B) > \frac{1}{2}P(A_n)$ and $P(A_n \bar{B}) > \frac{1}{2}P(A_n)$. Hence, let $A_{n+1} = A_n B$ if $P(A_n B) \leqslant \frac{1}{2}P(A_n)$; otherwise, let $A_{n+1} = A_n \bar{B}$. Then, because $A_{n+1} \subset A_n \subseteq A$, (i) is satisfied. Further, because $2P(A_{n+1}) \leqslant P(A_n) \leqslant P(A)/2^n$, (ii) is satisfied. This completes the induction.

Choose n such that $P(A)/2^n < b$. Then $0 < P(A_n) < b$, contradicting the assumption that $(0, b)$ is a gap within A. Q.E.D.

Lemma 2. For any A, if there is a gap within A, then A contains a gap of the form $(0, b)$.

Proof. Let (a, c) be a gap within A. Then, for any $B \subseteq A$, let the set of real numbers $S[B] = \{r \mid \text{for some } C \subseteq B, c \leqslant P(C) = r\}$. Let the number $s[B]$ be the greatest lower bound of $S[B]$. Thus, $s[B]$ is the greatest lower bound of the probability values for propositions within B that are at least as great as c. We now use induction to show that there is an infinite sequence of propositions $A_0, A_1, \ldots$ such that the following hold, for all i: (i) if $i > 0$, then $A_i \subseteq A_{i-1} \subseteq A$; (ii) $P(A_i) \geqslant c$; (iii) (a, c) is a gap within A_i; (iv) $P(A_i) - s[A_i] \leqslant 1/2^i$.

Base step: Let $A_0 = A$. Then (i)–(iv) are obviously satisfied.

Induction step: Suppose that A_n exists and satisfies the four conditions. Then $S[A_n]$ is nonempty, because it contains $P(A_n)$. Thus, $s[A_n]$ exists and is at least as great as c. Then for any $\varepsilon > 0$ there is a $B \subseteq A_n$ such that $s[A_n] \leqslant P(B) < s[A_n] + \varepsilon$ and $P(B) \geqslant c$. Choose $\varepsilon < 1/2^{n+1}$, and choose a $B \subseteq A_n$ accordingly. Let $A_{n+1} = B$. We verify that (i)–(iv) hold for A_{n+1}.

158

That (i) and (ii) hold is obvious; that (iii) holds follows immediately from (i), (ii), and the fact that (a, c) is a gap within A_n. Finally, we know that $P(A_{n+1}) - s[A_n] < 1/2^{n+1}$; so to establish (iv) we need only show that $s[A_{n+1}] \geq [A_n]$. So suppose that $s[A_{n+1}] < s[A_n]$. Then there is some $C \subseteq A_{n+1}$ such that $s[A_{n+1}] \leq P(C) < s[A_n]$ and $P(C) \geq c$. But because $A_{n+1} \subseteq A_n$, it follows that $C \subseteq A_n$, $C < s[A_n]$, and $P(C) \geq c$ – which contradicts the definition of $s[A_n]$. So there is no such C. Hence, in fact, $s[A_{n+1}] \geq s[A_n]$. This completes the induction.

The A_i form a nested sequence: For any $i, j, A_i \subseteq A_j$ if $j < i$. Let $B = \cap \{A_i\}$. Then $B \subseteq A$, and $P(B) = \lim_{i \to \infty} P(A_i) \geq c$. So $s[B]$ exists and is at least as great as c. Now, for any $i, s[B] \geq s[A_i]$. For suppose otherwise: Then there is some $C \subseteq B$ such that $s[B] \leq P(C) < s[A_i]$ and $P(C) \geq c$. But because $B \leq A_i$, it follows that $C \subseteq A_i$, $C < s[A_i]$, and $P(C) \geq c$ – which contradicts the definition of $s[A_i]$. So there is no such C. Hence, in fact, $s[B] \geq s[A_i]$, for all i.

Then $s[B] = P(B)$, for if not, then $s[B] < P(B)$, and this leads to a contradiction: If $s[B] < P(B)$, then choose an i such that $P(B) - s[B] > 1/2^i$. Because $s[B] \geq s[A_i]$ and $P(B) \leq P(A_i)$, it follows that $P(A_i) - s[A_i] > 1/2^i$ – which contradicts condition (iv).

Because $s[B] = P(B)$, there is within B a gap of the form $(a, P(B))$. Hence there is also a gap of the form $(0, b)$, where $b = P(B) - a$. Q.E.D.

Lemma 3. If no propositions in $\langle W, \mathsf{F}, P \rangle$ contain gaps, then $\langle W, \mathsf{F}, P \rangle$ is full.

Proof. Choose an arbitrary $A \in \mathsf{F}$ and $r \in (0, P(A))$ [we assume that $P(A) > 0$]. We need to show that there is some $B \subset A$ such that $P(B) = r$. We first use induction to show that there is an infinite sequence of propositions $B_0, B_1, \ldots$ such that the following hold, for all i: (i) $B_i \subset A$; (ii) for all $j \leq i$, $B_j \subseteq B_i$; (iii) $0 \leq r - P(B_i) < 1/2^i$.

Base step: Let $B_0 = \emptyset$. Then (i)–(iii) are obviously satisfied.

Induction step: Suppose B_n exists and satisfies (i)–(iii). If $r - P(B_n) < 1/2^{n+1}$, then let $B_{n+1} = B_n$, in which case (i)–(iii) are clearly satisfied. If not, then let $C = A - B_n$, and let $c = r - P(B_n)$. Because $r < P(A)$, $c < P(C)$. Let $a = c - 1/2^{n+1}$; then $a \geq 0$. Because C contains no gaps, it, in particular, contains no gaps of the form (a, c); hence there is some $D \subseteq C$ such that $a < P(D) < c$. Choose such a D, and let $B_{n+1} = B_n \cup D$. Then (i) and (ii) are obviously satisfied, as is (iii), for $c - P(D) < c - a = 1/2^{n+1}$; but also $c - P(D) = r - P(B_n) - P(D) = r - P(B_{n+1})$. This completes the induction.

Let $B = \cup \{B_i\}$. Thanks to (i), $B \subseteq A$. Thanks to (ii), $P(B) = \lim_{i \to \infty} P(B_i)$. Finally, thanks to (iii), $\lim_{i \to \infty} P(B_i) = r < P(A)$. Thus, $B \subset A$ and $P(B) = r$. Q.E.D.

159

The proof of the main result – that $\langle W, \mathsf{F}, P \rangle$ is full if atomless – follows immediately from Lemmas 1–3. Q.E.D.

NOTES

1. In particular, see Hájek and Hall (Chapter 6, this volume).
2. Why *conditional* – as opposed to absolute – probabilities? Only because there is a special case of a probability space $\langle W, \mathsf{F}, P \rangle$ with just four absolute probability values, but more conditional probability values: W contains three worlds, and each is given weight $\frac{1}{3}$. This probability space should not be considered trivial; indeed, it cannot be extended to a model for which CCCP holds.
3. Capital letters "A," "B," ..., "X," "Y,"... are to be understood to denote elements of F.
4. Suppose otherwise: Then for some $X \subset C$, $P(X) > P'(X)$. But then for some $i, P(XB_i) > P'(XB_i)$, contradicting the given. Note that, intuitively, C is the largest proposition over which P' dominates P.
5. Suppose not. Then, if C is not a P-atom, so that for some $D \subset C$ $0 < P(D) < P(C)$, we get a contradiction, as follows: Let $X = (D \cup \bar{C}) \to \bar{C}$. Using CCCP, $P'(X) = 0$ but $P(X) > P(\bar{C})$. Then $P(XC) > 0 = P'(XC)$ – the needed contradiction. So C is a P-atom. But that too yields a contradiction, because (as the next paragraph establishes) $\bar{C}$ is a P-atom, but P is nontrivial. Therefore $P'(\bar{C}) > 0$.
6. I thank Alan Hájek for drawing my attention to this equivalence, a proof of which is given in the Appendix.
7. Note that if (L2) held, we could stop here, because (iii), together with CCCP, violates it: $P(A \to B \cap A \to \bar{B}) \leqslant P(A \to (B \cap \bar{B}))$ by (L2); $P(A \to (B \cap \bar{B})) = 0$ by CCCP.
8. I harmlessly abuse conventions governing the use of parentheses in order to enhance readability.
9. Alan Hájek (1989) "Probabilities of Conditionals – Revisited." *Journal of Philosophical Logic* 18: 423–8.
10. It is still an open question, however, just what happens when the range of P is infinite but not equal to $[0, 1]$.

160

9

The Howson-Urbach proofs of Bayesian principles

CHARLES S. CHIHARA

Colin Howson and Peter Urbach write in the concluding paragraph of their book *Scientific Reasoning* (1989) that

> we want to demonstrate... that [Bayesianism] is the *only* theory which is adequate to the task of placing inductive inference on a sound foundation.

As part of this "demonstration," these authors present what they claim are "proofs" of the fundamental principles of Bayesianism. Their alleged proofs will be examined in this chapter.

1. THE COHERENCE PROOF

The first "proof" I shall discuss is one that has as its conclusion the Bayesian doctrine that an agent's degrees of belief ought to be "coherent" (i.e., obey the axioms of the probability calculus). In other words, an agent's degrees of belief ought to be given by a *probability function*. What is a probability function? Let S be a set of sentences containing at least one *tautology*, where a *tautology* is any sentence that is "true independently of the state of the world" (i.e., is a necessary statement). A probability function on S is a function that assigns non-negative real numbers to the members of S in such a way that every tautology is assigned 1, and the number assigned to a mutually exclusive disjunction is equal to the sum of the numbers assigned to the disjuncts. It needs to be noted that, according to these authors, two sentences are mutually exclusive if and only if "one entails the negation of the other" (Howson and Urbach, 1989, p. 16), where entailment is not restricted to logical entailment, but is "to be understood as incorporating all of contemporary mathematics" (p. 15). We can summarize the foregoing by means of axioms, as follows: A prob-

161

ability function P must be such that

1. $P(\sigma)$ is greater than or equal to zero, for all $\sigma \in S$,
2. $P(\sigma) = 1$ if σ is a tautology in S,
3. $P(\sigma \vee \tau) = P(\sigma) + P(\tau)$ if σ and τ are mutually exclusive sentences in S and $(\sigma \vee \tau)$ is a member of S.

(A function that satisfies axiom 3 is said to be "additive.")

What, according to these authors, are an agent's *degrees of belief*? Better yet, what does it mean to say that someone's degree of belief in sentence σ is given by real number r? They tell us that r reflects the extent to which the agent believes it likely that σ will turn out to be true. However, to see more specifically what degrees of belief are supposed to be, we need to understand what *subjectively fair odds* are.

Definition. Let $p/(1-p)$ be "those odds on a hypothesis h which, so far as you can tell, would confer no positive advantage or disadvantage to anyone betting on, rather than against, h at those odds.... Such odds, *if you can determine them*, we shall call your *subjectively fair* odds on h." (Howson and Urbach, 1989, p. 56; italics added)

Our degree of belief in σ, however, is not given by our subjectively fair odds on σ, but rather by another number called "the betting quotient associated with the fair odds." If $p/(1-p)$ describes our subjectively fair odds on σ, then p is the *betting quotient associated with σ*. Thus, to say that our degree of belief in σ is p is to say that our subjectively fair odds on σ are given by $p/(1-p)$.

The betting situations that these authors have in mind are given the following canonical form: There are exactly two bettors: one *betting on* the truth of the sentence σ, and the other *betting against* σ. There is a stake S that is given as so many units of currency, the units being assumed to be infinitely divisible. The betting quotient is a rational number p. If σ is determined to be true, the bettor on receives $S(1-p)$ units of currency from the bettor against. If σ is determined to be false, the bettor against receives Sp units of currency from the bettor on. Hence, a "payoff table" can be constructed for the bet on σ in the following way:

σ	Payoff
T	$S(1-p)$
F	$-Sp$

The minus sign, of course, indicates that the bettor on loses the sum of Sp when it is determined that σ is false.

It needs to be emphasized that for Howson and Urbach, degrees of belief are defined in terms of what the agent judges (or would judge) to be

162

subjectively fair odds, and not, as is quite frequently the case, in terms of bets that the agent accepts: "backing up judgments with financial commitment is a luxury not everyone can afford, even if they wanted to" (1989, p. 58). They emphasize that they do not assume or presuppose any behavioral consequences of having a degree of belief in some proposition.

How do these authors attempt to prove that an agent's degrees of belief should be given by a probability function? They attempt to show that betting quotients that do not satisfy the probability axioms cannot consistently be regarded as determining fair odds (1989, p. 59). Thus, consider what these authors say about a betting quotient that does not satisfy axiom 2. If τ is a tautology, and $P(\tau) = p > 1$, then from the preceding payoff table it can be determined that the bettor on τ will necessarily lose a sum of currency: Because τ is a tautology, we need only consider the row in the table beginning with T, and because $p > 1$, the payoff must be negative. But this is demonstrably unfair, because a positive advantage will be possessed by the bettor against (1989, p. 60). On the other hand, if $P(\tau) = p < 1$, then the bettor on τ will necessarily possess the positive advantage. In either case, the betting quotient cannot be fair.

Let us consider now the case in which an agent's degrees of belief do not satisfy axiom 3. We suppose this time that $P(\sigma) = p$, that $P(\tau) = q$, and that σ and τ are mutually exclusive. Suppose that the agent places two simultaneous bets, one on σ and the other on τ, each with the same stake S. The combined payoff table will look like this:

σ	τ	Payoff
T	F	$S(1 - p) - Sq$
F	T	$-Sp + S(1 - q)$
F	F	$-Sp - Sq$

Noticing that the payoff for the T–F row is identical with the payoff for the F–T row, it can be seen that this table is also the payoff table for the bet on $(\sigma \vee \tau)$ with betting quotient $(p + q)$ and stake S.

Suppose now that the agent's degree of belief in $(\sigma \vee \tau) = r$ and that $r \neq p + q$. Then the agent will consider a bet on $(\sigma \vee \tau)$ to be fair if it has a payoff table that differs from the preceding in the payoff column: Its payoff column will have to differ from the preceding because $r \neq p + q$. Thus, it is argued, if the agent's degree of belief in $(\sigma \vee \tau) \neq p + q$, then the foregoing considerations show that the agent will consider bets to be fair if they entail two distinct odds being given on $(\sigma \vee \tau)$. They conclude:

The situation is thus analogous to an inconsistent set of sentences: the attempt to assign them all the truth-value 'true' results in one of the sentences being assigned different truth-values. In the present case the assignment is not of truth-values, but betting quotients, but the inconsistency is no less. [Howson and Urbach, 1989, p. 61][1]

The conclusion arrived at is this: Degrees of belief must be additive (i.e., must satisfy axiom 3) on pain of inconsistency. I shall analyze this alleged proof shortly. But first, let us consider their proof of a more controversial Bayesian principle.

2. The Principle of Conditionalization

Another Bayesian principle that Howson and Urbach claim to prove is the Principle of Conditionalization, which they state in several different ways. The principle is first mentioned in Chapter 3, where it is described as "the foundation of the theory of Bayesian inference" (Howson and Urbach, 1989, p. 67). Then they give a rough and partial statement of the principle, writing: "What it is necessary to show to exploit Bayes's Theorem in this way is that $P'(h) = P(h/e)$, where $P'(h)$ is your degree of belief in h after receipt of e" (1989, p. 68). Much later, in Chapter 11, they give the following characterization of this principle:

[C] We are proposing a theory of inference; in particular, a theory in which from two inputs, e and a belief distribution, an output belief-distribution is generated by the Principle of Conditionalisation. [1989, p. 285]

What, then, is this principle that is the basis for the Bayesian theory of inference? Here is how they state it in Chapter 11:

[PC-1] [I]f $P(h/e)$ is your conditional probability of h on e, and you learn e (but nothing stronger), then consequent upon this information, your degree of belief in h is, if you are consistent, equal to $P(h/e)$. [1989, p. 258]

Later, however, they give it in a significantly different way:

[PC-2] Somebody who had degrees of belief $P(h)$, $P(e)$, and $P(h\&e)$ in the truth of the sentences h, e, and $h\&e$ thereby has, on pain of inconsistency,... a degree of belief $P(h/e)$ in h, conditional on e's being true, where $P(h/e) = P(h\&e)/P(e)$. If e does turn out to be true, then the degree of belief of this person, again on pain of inconsistency, in h unconditionally becomes $P'(h) = P(h/e)$. [1989, p. 284]

What is a *degree of belief in a sentence that is conditional on some sentence*? The authors do not, as is usual, define $P(\sigma/\tau)$ to be the ratio

$$P(\sigma \& \tau)/P(\tau)$$

where $P(\tau) \neq 0$. Instead, they define an agent's conditional degree of belief in σ, given τ, in terms of a certain conditional bet on σ. But first we need to see how they specify what a conditional bet is. A conditional bet on σ, given τ, with betting quotient q, goes ahead with that betting quotient

164

"on receipt of an acceptable affidavit" of the truth of τ, and is called off on receipt of an acceptable affidavit of the falsity of τ (1989, p. 63). An agent's conditional degree of belief in σ, given τ, is p if and only if the agent *would deem fair a conditional bet on σ, given τ*, with betting quotient p. They then specify that the notation "$P(\sigma/\tau)$" is to be used to refer to the agent's conditional degree of belief in σ, given τ (1989, p. 63).

Let us now consider their proof of the Principle of Conditionalization. This proof is given long before the principle in question is even stated. It is claimed in Chapter 11 that the validity of the principle had been proved in Chapter 3. So we shall examine the proof given there. The first part of their proof consists in an argument to show that for any sentences σ and τ, $P(\sigma/\tau) = P(\sigma \& \tau)/P(\tau)$. Briefly, they argue that if $P(\sigma \& \tau) = q$ and $P(\tau) = r$, then if $P(\sigma/\tau) \neq q/r$, the agent "would implicitly be assigning different odds to the same (conditional) hypothesis" (1989, p. 65). The reasoning here is similar to their reasoning to show that degrees of belief must be additive.

The second part of this proof attempts to show that $P'(h) = P(h/e)$, where $P'(h)$ is "your degree of belief in h after receipt of e" (1989, p. 68). Because this reasoning will be examined in detail, I quote the relevant passage:

If, as will be assumed, the background information relative to which $P'(h)$ is defined differs from that to which $P(h/e)$, $P(h)$, $P(e)$, etc., are relativised only by the addition of e, the principle follows immediately; for $P(h/e)$ is, as far as you are concerned, just what the fair betting-quotient would be on h were e to be accepted as true. Hence from the knowledge that e is true you should infer (and it is an inference endorsed by the standard analysis of subjunctive conditionals) that the fair betting quotient on h is equal to $P(h/e)$. But the fair betting quotient on h after e is known is by definition $P'(h)$. [1989, p. 68]

This reasoning has some very strange elements. First of all, there is the assumption about "background information." What is background information? Is this supposed to be "information" that the agent has at the time her degrees of belief are "$P(h/e)$, $P(h)$, $P(e)$, etc."? Or are they merely degrees of belief that she has at that time? What turns degrees of belief into "information"? Must the degree of belief in e be 1 before it counts as "information"? Must the belief in e be acquired in some way before it counts as "information"? What makes information "background" (as opposed to "foreground")? But more importantly, what allows them to *assume* that "the background information relative to which $P'(h)$ is defined differs from that to which $P(h/e)$," and so forth, are relativized *only by the addition of e*? Reread the two statements of [PC] again. First of all, there is nothing there about "background information." Second, [PC-2] does not even imply that there is any change of information that takes place; it only has the condition that e turns out to be true. On the other hand, [PC-1] does imply that there is a change of information to be considered,

165

insofar as it says that you learn e. But it does not say that e is the only thing that is learned; it only says that you learn "nothing stronger" (whatever that means). If they are simply making a gratuitous assumption here, then this assumption will have to be discharged before one will have an acceptable proof of the principle.

Besides, there is something very strange about this assumption they make. How could the background information that the agent has at the later time differ from what she had at the earlier time *only by the addition of e*? If she has received information e, she will of necessity have received it in some way (e.g., by direct observation, by experimentation, by reading about it in a journal, or by obtaining e from a reliable source). Furthermore, she will, no doubt, also realize *that she has been given this information*. She will also, very likely, realize that this additional information has affected her degrees of belief in all sorts of things and in all sorts of ways. All of this will introduce into her fund of "background information" sentences – it would seem, a large number of additional sentences – and in some cases this additional information may be crucial (depending upon what h is).

This reasoning is also dubious because of the remarkable shifts that Howson and Urbach make, with not a word of justification. Notice that in their statement of the principle [PC-2], they do not require that e be accepted as true, or come to be known to be true, or be received as information: They only require that e *turn out to be true*. Yet they feel justified in assuming that e is added to the agent's background information. Could not e turn out to be true without it being added to the agent's background information? Could not e be true even though the agent is not informed about it? They then slip into speaking as if the agent has accepted e as true. And finally they speak as if the agent has knowledge that e is true. All of this is decidedly un-Bayesian. What does it mean to *accept e* as true? In his response to Isaac Levi's paper (1970) criticizing Bayesianism, Richard Jeffrey (1970, p. 183) wrote: "[M]y dissatisfaction with Levi's positive proposals ... was centered on the very notions of acceptance and rejection. ... [W]hile he gave methods for deciding which (if either) of the two acts *accept H* and *reject H* one ought to perform, he provided no account of how one is to go about performing those acts." To the Bayesian Jeffrey, such talk of acceptance is unclear, whereas talk of degrees of belief is a model of clarity. It is thus interesting to see Howson and Urbach shifting from talk about degrees of belief to talk about accepting e as true. Then there is the other shift from talk of accepting e to talk about knowledge of e. Do they believe that acceptance of a sentence as true amounts to having knowledge of the truth of this sentence? What is going

166

on here? All of the foregoing points indicate a shocking lack of rigor to be found in the alleged "proof" of [PC].

Generally, Bayesians do not state the Principle of Conditionalization in the way Howson and Urbach do. For example, as Jeffrey (1983, p. 164) describes the principle in his well-known textbook, to apply the principle it is not required that e turn out to be true, as Howson and Urbach imply with [PC-2]; instead, it must be the case that the agent's degree of belief in e change to 1 [i.e., that $P'(e) = 1$].[2] Thus, it can be seen that as Howson and Urbach state the principle [PC-2], they have replaced the usual sub-jective condition that the agent becomes certain of the truth of e with the objective condition that e be true. Let us suppose, however, that these authors have simply made a slip here and that we take [PC-1] to be the more accurate statement of the principle they have in mind. In an effort to make their proof more cogent, I shall suppose in what follows that they had meant to state the principle using the subjective condition that the agent has "learned" e or has "accepted" e as true (in some sense of these terms).[3]

As Jeffrey states the principle, there is no need to bring in such murky notions as *acceptance* or *knowledge*. Why, then, do Howson and Urbach resort to such sloppy talk? I can only conjecture here, but one possibility is that they were not comfortable with the more standard account of the principle, especially given the heavy use they wished to make of the principle in accounting for scientific reasoning. How often is a scientist's degree of belief in an evidential statement so high as to correspond to the number 1? Practically never, I should think. Typically, an evidential statement that is thought to confirm or disconfirm some scientific hypothesis is asserted as a result of an experiment of some sort that depends upon a number of factors for its accuracy and correctness: the accuracy of adjust-ments and calibrations of various instruments; the proper functioning of measuring devices, machines, and instruments; the purity and exact propor-tions of chemical solutions; the accuracy of an auxiliary hypothesis – these are just some of the factors that may be presupposed. In a huge number of cases, scientists draw conclusions from evidential statements that do not warrant the kind of degrees of belief that the standard form of the Principle of Conditionalization requires. So it would have been natural for Howson and Urbach to have stated the principle either in a way that did not require any such subjective certainty or in a way that would obscure that weakness. Alas, the result is a number of very questionable transitions in their reasoning.

Howson and Urbach try to do a great many things in their book *Scientific Reasoning*. Besides trying to prove the principles of Bayesian subjective probability theory, they also attempt to show how these principles can be applied in the justification and analysis of inductive reasoning of all sorts, including scientific and statistical reasoning. They also attempt to answer the chief objections that have been raised to Bayesian confirmation theory. I believe that some of the nonstandard features of their versions of the Bayesian principles arise out of their attempts to answer these objections.

Consider, for example, their way of responding to the Old Evidence Problem. As Clark Glymour (1980, pp. 85–93) has pointed out,[4] some scientific theories have received support from data that were gathered before the theories were formulated. Glymour cites such cases as the support that Copernicus' theory received from astronomical observations made much earlier. Because these old data had already been obtained,

$P(e)$ is 1, whence $P(e/h)$ is 1 also, so that it follows immediately from Bayes's Theorem that $P(h/e) = P(h)$. Thus e does not raise the prior probability of h and hence, according to the Bayesian, does not confirm it. [Howson and Urbach, 1989, p. 270]

Glymour's Old Evidence Problem is aimed at what Howson and Urbach characterize as "the foundation of the Bayesian theory of inference," namely, the Principle of Conditionalization and, more specifically, the Bayesian kinematic picture, which Glymour describes as follows:

[A] Bayesian agent moves along in time having at each moment a coherent set of degrees of belief; at discrete intervals he learns new facts and each time he learns a new fact, e, he revises his degrees of belief by conditionalizing on e. [Glymour, 1980, p. 83]

In the foregoing case, because $P(h/e) = P(h)$, there does not seem to be any ground for raising one's degree of belief in h by means of the Principle of Conditionalization. However, it can be seen that the realization that the hypothesis h explains the old evidence e convinces the agent that this evidence confirms the hypothesis; and this generates an increase in the agent's degree of belief in h.

Howson and Urbach respond to this objection by claiming that the Bayesian inductive principle has not been correctly applied. They claim that "the mistake lies in relativising all the probabilities to the *totality* of current knowledge: they should have been relativised to current knowledge minus e" (1989, p. 271). In determining if the old data e should raise one's degree of belief in h, one should not evaluate the probabilities in terms of the knowledge state in which the agent knows e. They claim that "the support of h by e is gauged according to the effect which a knowledge of e *would* now have on one's degree of belief in h, on the (counterfactual) supposition that one does not yet know e" (1989, p. 274).

168

I cannot help but wonder how these authors can be comfortable with such a reply. First of all, there is nothing wrong with Glymour's applications of the Principle of Conditionalization. His applications of the principle accord perfectly with standard versions that Bayesians have accepted for years. Indeed, these applications conform to the versions [PC-1] and [PC-2] that Howson and Urbach themselves give. It is true that Glymour's applications do not conform to the version of the principle that Howson and Urbach put forward in their reply. But that version is not the version that they attempt to prove. Notice that the version of [PC] that was discussed in the preceding section makes no mention of assessing degrees of belief in any kind of counterfactual knowledge states. What, then, justifies their appeal to this new version of [PC]? Perhaps they believed that the version of [PC] to which they appealed in responding to the Old Evidence Problem was a mere variation of their more standard [PC]. Because the more standard version had been justified, so they thought, the variation was also. Of course, one need only express the reasoning to see how flawed it is. Howson and Urbach seem to be indulging in a kind of shell game. They give one version of the principle when attempting to prove the principle. But then they give a different version when responding to objections.

Why did they not simply state and prove the version of [PC] they wished to use in their reply? I suspect that they felt that any version they could have formulated that could have been used to back their reply to the Old Evidence Problem either would have sounded absurd or would have been unprovable. Consider the difficulty of stating the principle. One is supposed to relativize probabilities to "current knowledge minus e." But it clearly will not do to eliminate only e from "current knowledge": If one leaves in sentences that are logically equivalent to e, one will still obtain the counterintuitive results that Glymour has described. But eliminating all sentences logically equivalent to e will not be sufficient, for e may be necessarily or mathematically equivalent to other sentences in "current knowledge," and such sentences will still raise Glymour's problem. In fact, e may be theoretically equivalent to other sentences in "current knowledge" (i.e., e may be equivalent, within some accepted scientific theory, to other sentences in "current knowledge"), and such sentences will raise problems too. I suppose that Howson and Urbach would also want to delete all sentences that theoretically imply e. But what about sentences that do not theoretically imply e, but merely make e probable?

Consider the following: Suppose that for a number of years it has not been realized that some well-known experimental result β confirms some hypothesis σ. When it is finally observed that β does confirm σ, we run into the Old Evidence Problem. So clearly we need to relativize our probabilities to "current knowledge" minus a lot of sentences. Now even if we subtract all sentences that logically, mathematically, or theoretically imply β, there will still remain in "current knowledge" many sentences that would not be there had the result expressed by β not been obtained. And some of these sentences may affect the probability of β and σ. (One of these could be the sentence that asserts that Professor X, a careful and cautious researcher, believes β.) Should we then delete all sentences that affect the probability of β and σ? But that clearly would go too far.

Besides the problem of stating just what sentences we are supposed to delete from "current knowledge" in making our assessments of fair betting quotients, there is also the formidable problem of actually making the assessments. Thus, returning to the foregoing example, the experiments that gave rise to β may have affected, in countless ways, both far-reaching and subtle, the degrees of belief we have in the truth of all sorts of sentences that may be relevant to our degrees of belief in σ, β, and $(\sigma \& \beta)$. So are we supposed to calculate what our degrees of belief would have been had these experiments not been conducted? And why should we suppose that one can, in any reasonably accurate manner, carry out such calculations? And if one cannot carry out such calculations, how can we be sure that one would consider any conditional bet on σ, given β, a fair bet? How can we *prove* that any agent would consider some conditional bet on σ, given β, a fair bet? Perhaps, then, the relevant $P(\sigma/\beta)$ would not even exist. (Recall the definition of subjectively fair odds.) And if we cannot be sure of that, then what happens to our principle?[5]

There are some things Howson and Urbach have said that suggest a kind of reply to the sort of objection I have been making. The suggestion is that I am merely pointing to a practical problem – the problem of determining to what specific body of background information an agent should relativize her degrees of belief $P(h/e)$ and $P(h)$. This practical problem does not undermine the theoretical underpinnings of Bayesian confirmation theory, it might be thought, because "people are capable in many cases of determining, possibly only very roughly, to what extent they think a piece of data likely relative to a stock of residual background information" (1989, p. 273).

But such a reply misses the point of the objection I am raising. Glymour's Old Evidence Problem is aimed at the Principle of Conditionalization. Howson and Urbach respond to Glymour's problem by, in effect, appealing to some other principle of subjective probability change. The problem I am raising to the Howson-Urbach response is not a practical problem at all – it is the theoretical problem of stating this principle of probability change. Before they can claim to *prove* such a principle (which they seem to think they have already done), they need, first of all, to state the principle.

170

Not only have they not stated the principle; it is even questionable that they *could* state it in a satisfactory way. This, at any rate, is what I have been arguing.

There is another problem that must be resolved before we can have anything like a clear statement of an acceptable principle of subjective probability change. Recall the characterization [C] of the Principle of Conditionalization that Howson and Urbach give. An agent has various degrees of belief, among which are $P(h/e)$ and $P(h)$. (This is the initial belief distribution.) She has accepted e, which, we can assume, for now, means that $P(e) = 1$. (Thus, we have the two inputs described in the characterization.) It is only some time later that the agent realizes that h explains e. What should her degree of belief in h be as a result of this realization? In short, what should the "output distribution" be that is generated by this principle? The principle of probability change under discussion tells us, I assume, that the agent is supposed to relativize her subjective probabilities to "current knowledge" minus e (and presumably everything else dependent upon e). Assuming that this can be done and that we get subjective probabilities $P^*(h/e)$ and $P^*(h)$, relativized to this body of background information, what are we supposed to do with these degrees of belief? We still have not been told what the agent's new degree of belief in h should be. Should it be $P(h) + (P^*(h/e) - P^*(e))$? If so, why? Howson and Urbach give us no idea at all, nor is there anything in the book that is even remotely like a proof that such an output distribution is required. So we have no clear statement of any principle of subjective probability change to assess for plausibility. And insofar as they have given, in their book, what they consider to be a proof of a principle of probability change, this supposed proof is not a proof of the principle they need in order to respond to Glymour's Old Evidence Problem.

3. THE FATAL FLAW

I shall now take up what I take to be the fatal flaw in the Howson–Urbach proof of [PC]. One thing I would like to emphasize in my analysis that these authors did not is the time factor. As many philosophers have noted, the Principle of Conditionalization is a dynamic principle. We suppose that h is some hypothesis and that e is some evidential statement bearing on the truth of h. (Howson and Urbach suggest that we take e to be "the result of some test.") We also suppose that, at some time t, you have a coherent set of degrees of belief given by probability function P. At some later time t^*, you "learn" e (in some sense of that term). [PC] tells us that your degree of belief in h at t^*, $P'(h)$, ought to be $P(h/e)$. According to Howson and Urbach, the reason $P'(h)$ ought to equal $P(h/e)$ is this: "$P(h/e)$

is, as far as you are concerned, just what the fair betting-quotient would be on h were e to be accepted as true." But if we make the time factor explicit, what they are claiming is this: $P(h/e)$ is, as far as you were concerned at time t, just what the fair betting quotient would be on h were e to be accepted as true. It does not logically follow that, at time t^*, $P(h/e)$ is, as far as you are concerned, just what the fair betting quotient would be on h were e to be accepted as true. In short, there is a lacuna in their proof: They have simply assumed (without the slightest justification) that if you have not received any information between time t and time t^* other than e, then all your relevant degrees of belief should remain the same. But why should that be so? What principle of rationality demands any such conservation of degrees of belief?

Suppose that h is this sentence: "The Oakland A's win the American League Pennant in 1991." And suppose that e is this sentence: "The A's are leading the league on July 4, 1991." Suppose further that on January 1, 1991, your conditional degree of belief in h, given e, is $P(h/e)$. Now if you do not obtain any information relevant to h during the month of January, does it follow that your conditional degree of belief in h, given e, should remain the same? Might you not come to think in late January that your earlier assessment of Oakland's chances of winning the pennant was too high? Perhaps you come to think that the psychological makeup of some of the stars of the team may be harmful to the team's spirit. Or perhaps you begin to worry about the age of some of the key players. Have Howson and Urbach justified any principle of rationality that precludes an agent from making such a change in degrees of belief "on pain of inconsistency"? Not at all. Their alleged proof has a huge hole in it.[6]

I can imagine someone attempting to solve this problem by adding a clause to the statement of the principle: The revised principle would state in the antecedent something to the effect that none of the agent's relevant degrees of belief change between time t and time t^*. But such a revision will not do the trick. The acceptance of e may itself bring about an insight that results in a radical change in the agent's degrees of belief. In an earlier paper, I described such a case: A prince is given the task of guessing the number of a ball that he is to be given on the Saturday after having received a numbered crimson ball on each of the previous six days of the week. Upon receiving the sixth ball, he suddenly realizes that the six numbers he has received are all Gödel numbers (according to the numbering system of the king's logic book) and that they are, in the order received, the numbers of the letters

Crimso

The prince's degree of belief in the hypothesis that the seventh ball will be numbered "69" – the Gödel number of the letter "n" – goes way up. And this jump in probability is not the result of conditionalizing on the evidential statements. So even if the prince's degrees of belief had not changed until after he received the sixth ball, it would be absurd to insist that he stick with the degrees of belief he had at the beginning of his test and merely conditionalize on the sentences that express the data he received via the balls.[7]

At one time, there was a tendency among Bayesians to think that a change of degrees of belief that did not conform to the Principle of Conditionalization was simply irrational or not rational. It should be abundantly clear that, in this case, the change in the prince's degrees of belief that took place as a result of noticing that the numbers he had received were Gödel numbers was definitely rational. Certainly it would be more rational to change one's degrees of belief in the way the prince did than, stubbornly, to cling to his Bayesian principles and change his degrees of belief only by conditionalization. The Howson-Urbach "proof" that an agent must infer, at t^*, that the fair betting quotient on h, after receiving e, is equal to the conditional degree of belief that she had at t on h, given e, simply does not hold up to scrutiny.

There is an obvious connection between this example of a change of belief brought about by insight and Glymour's Old Evidence Problem. In Glymour's examples, too, each change in partial beliefs is brought about by an insight: the realization that the old evidence is implied by the hypothesis. The Bayesian principles are simply not sensitive to changes of this sort. A major source of the difficulty is the very unrealistic treatment of agents as if they knew all necessary truths. This point becomes crucial in the following analysis of the alleged proof that a person's degrees of belief must satisfy the axioms of probability.

4. AN ANALYSIS OF THE COHERENCE PROOF

Howson and Urbach claim to prove that an agent's degrees of belief must satisfy the axioms of probability "on pain of inconsistency." Indeed, they claim to show that the rules of probabilistic reasoning "are broken on pain of committing inconsistency" (1989, p. 295). But how could they have proved any such thing? It seems evident that a perfectly reasonable agent could have degrees of belief that do not satisfy those axioms. Recall that by "tautology," Howson and Urbach mean "necessary truth." Now no perfectly reasonable human being should be expected to have degrees of belief that satisfy axiom 2. For example, no living human knows whether or not Fermat's Last Theorem (henceforth F) is true. Hence, no human

173

being, even a perfectly reasonable one, should be expected to have a degree of belief in F equal to 1 or 0.

Near the end of their book, Howson and Urbach respond to an objection in a way that may suggest to some readers that they have a reply to the foregoing objections. They do not claim, they say, that everybody ought to be consistent, nor that reasonable people are always consistent. Then they go on to say:

> The foundation of subjective Bayesianism is probabilistic consistency, admittedly, but this is an ideal which is not always attainable.... *All we claim is that when people recognise or are apprised of deductive relationships between hypothesis and evidence, they often draw conclusions about levels of support in accordance with those determined within the Bayesian theory by the same initial probabilities.* [1989, pp. 273–4; italics added]

But is *this* all they are claiming when they advance their probabilistic consistency thesis? Is the subjective Bayesian theory of inductive reasoning to be based on no more than this? How can that very weak claim possibly be sufficient as the basis for the Bayesian theory of inference? The claim is so weak as to be almost insignificant: *People often draw conclusions about levels of support in the Bayesian way*? People often draw conclusions about levels of support in non-Bayesian ways too. People often do all sorts of silly things. What can we conclude from that about what *correct* inductive reasoning is?

I am afraid that what we see here is an expression of the unfortunate tendency of Howson and Urbach to describe their position in one way when they are laying the foundation for their theory of inductive inference, but to describe their position in quite different way when they are responding to objections. In many places, Howson and Urbach make much stronger claims about the relationship between degrees of belief and the axioms of probability. In particular, they claim that an agent's degrees of belief ought to conform to the rule of the probability calculus. For example, in the Introduction, they tell us that in Chapter 3 they will explain why "it is reasonable to expect that our degrees of belief in hypotheses should obey the axioms of probability" (1989, p. 12). Section D.2 of Chapter 3 is entitled "Why Should Degrees of Belief Satisfy the Probability Calculus?" In discussing the Principle of Conditionalization, they claim that "from the knowledge that e is true you should infer. ..." (1989, p. 68). And after discussing a number of arguments for the conclusion that subjective probabilities should obey the standard probability axioms, they write:

> [N]one seems to provide any conclusive reason for asserting that degrees of belief ought to obey the probability calculus. ... However, they are far from valueless.

It is a striking fact that, starting from often apparently very different assumptions, all plausible in their own way, so many arguments lead directly to the probability calculus. ... But enough, we feel, has been said about why degrees of belief should be formally probabilities. [1989, pp. 75–6]

These normative claims form the foundation of the Howson–Urbach Bayesian theory of inductive inference, and it is these claims that I am questioning. Evidently, Howson and Urbach believed that these normative claims could be justified by means of the following principle, which they thought they could prove:

[FB] Degrees of belief that do not conform to the probability calculus correspond to judgements of subjectively fair bets that cannot be fair.

From [FB], it could be argued that an agent with degrees of belief that did not satisfy the axioms of probability would have judgments about the fairness of bets that were demonstrably wrong. So, it could be argued, an agent should attempt to form degrees of belief that would conform to the probability calculus.

That Howson and Urbach reasoned in this way is strongly suggested by the following: In Section D.2 of Chapter 3, the authors claim to prove the central result of the chapter: "[I]f a set of betting quotients fails to satisfy the probability calculus, then were anybody to bet indifferently on or against the associated hypotheses, at the odds determined by those quotients, he or she could be made to suffer a net loss (or gain) independently of the truth or falsity of those hypotheses" (1989, p. 59). The importance of this result, they tell us, lies in the corollary that betting quotients that do not satisfy the probability calculus "cannot consistently be regarded as determining [subjectively] fair odds" (1989, p. 59). The crucial idea behind this corollary is this: If a person can, "on the basis of an examination of the odds alone, be assured of a positive net gain or loss from simultaneous bets at those odds, then the net advantage in betting at those odds cannot be zero" (1989, p. 59). They go on to conclude:

Thus axioms 1 and 2 of the probability calculus are necessary conditions for fairness: were these axioms not satisfied, then as the reasoning above shows, anybody with the equivalent degrees of belief, measured in the way we have measured them, would be classifying certain types of bet as [subjectively] fair which are demonstrably not. [1989, p. 61]

And later, they assert: "We have now completed the proof that if a set of betting quotients do [sic] not satisfy the probability calculus, then they certainly cannot all be fair" (1989, p. 65).

To obtain a sound assessment of this reasoning, let us ponder the argument these two authors construct to show that an agent's degrees of belief ought to satisfy axiom 3. Recall that a crucial step in the argument

175

is the presentation of the payoff table for simultaneous bets on σ and τ. Notice that the row for the case in which σ is true and τ is true has been left out. Why have they left that row out? Obviously, it is because that case is an impossibility: σ and τ are mutually exclusive. But suppose, as might well be the case, that the fact that σ and τ are mutually exclusive is unknown to every human being. After all, humans know only a very small number of necessary truths – as mathematicians would put it: "Humans know *almost no* necessary truths." From the point of view of the agent, then, both σ and τ could be true. Howson and Urbach have constructed the payoff table from the point of view of *a being who knows all necessary truths*. But an agent's degrees of belief are determined from the point of view of *the agent*, that is, by what the agent thinks are fair odds, from what she knows, what she can determine, what she can figure out. (Here, the reader may wish to review the definition of "subjectively fair odds" that was given earlier.) Thus, from her point of view, the T–T row may be a real possibility and should be left in. Thus, there may be no inconsistency *from her point of view* in not having a degree of belief in $(\sigma \vee \tau) = P(\sigma) + P(\tau)$. Indeed, there may be no inconsistency, from the point of view of *all* humans, in not having a degree of belief in $(\sigma \vee \tau) = P(\sigma) + P(\tau)$, and hence there may be no inconsistency, from the human point of view, in having degrees of belief that do not satisfy axiom 3.

To reinforce this last point, let us consider the following train of thought. We start with a few definitions. Sentences σ and τ are *mutually exclusive if and only if there is some true physical law that precludes σ and τ both being true. A probability function P satisfies axiom 3* (and is *additive) if and only if for all sentences σ and τ, if σ and τ are *mutually exclusive, then $P(\sigma \vee \tau) = P(\sigma) + P(\tau)$. It is now argued that P must satisfy axiom 3* on pain of inconsistency. Suppose that $P(\sigma) = p$ and $P(\tau) = q$, and suppose that the agent places two simultaneous bets, one on σ and the other on τ, each with the same stake S. The payoff table will look like this:

σ	τ	Payoff
T	F	$S(1-p) - Sq$
F	T	$-Sp + S(1-q)$
F	F	$-Sp - Sq$

It can be seen that this table is also the payoff table for the bet on $(\sigma \vee \tau)$ with betting quotient $(p + q)$ and stake S. Thus, it is argued, if the agent's degree of belief in $(\sigma \vee \tau) \neq p + q$, then the foregoing simultaneous bets on σ and τ will entail two distinct odds being given on $(\sigma \vee \tau)$.

What is wrong with this reasoning? Here again, it is unreasonable to eliminate the T–T row from the payoff table. Suppose that the agent does

176

not know that σ and τ are *mutually exclusive because she does not know of the existence of the relevant true physical law. From the agent's point of view, the foregoing table does not accurately describe the payoff. And it is unreasonable to condemn the agent's degrees of belief as entailing a commitment to subjectively unfair bets on the basis of a table constructed from the point of view of an omniscient being. From the subject's point of view, there is nothing unfair about the bets that correspond to the subject's degrees of belief: If the agent does not know of the existence of the relevant physical law, then bets made in accordance with her degrees of belief are not unfair, from her point of view. That is, the odds she gives are *subjectively fair odds*. Similarly, in the Howson–Urbach discussion of the example relating to axiom 3, the agent's degrees of belief are being condemned as leading to subjectively unfair bets because of the existence of a necessary truth that no human may be aware of. But, as in the foregoing case, from the subject's point of view, there may be nothing unfair about the bets that correspond to her degrees of belief: If she does not know of the existence of the relevant necessary truth, then those bets can be *subjectively fair*.[8]

Contrary to what Howson and Urbach claim, I have been suggesting that the Bayesian principles do not correctly state how the degrees of belief of real humans ought to be interrelated. As I see it, *the Bayesian principles are neither descriptively adequate nor normatively correct*. Hence, any attempt to "prove these principles" is doomed to failure. Instead of attempting to prove these principles, Bayesians should be trying to determine in what sorts of situations those unrealistic idealizations and presuppositions of their principles are relatively harmless. How much better it would be to seek improvements of the theory than to give bad proofs of incorrect principles.[9]

NOTES

1. Howson and Urbach inform us that they owe much of their proof to Chapter 6 of Brian Skyrms' *Choice and Chance* (1977). Skyrms (1987) suggests that this reasoning goes back to F. P. Ramsey, who had argued that if a person's degrees of belief did not satisfy the axioms of probability theory, she would be guilty of a kind of logical inconsistency.
2. Ellery Eells (1982, pp. 13–14) gives such a characterization.
3. This way of interpreting their version of the principle would fit the rough and partial statement of the principle (mentioned earlier) that these authors give just before the final part of their proof (Howson and Urbach, 1989, p. 68).
4. For a more detailed discussion of the Old Evidence Problem, and also for additional references, see Chihara (1987).
5. Points similar to the foregoing were made earlier (Chihara, 1987).
6. The gap noted here is well known. Thus, Hugh Mellor wrote: "Just as, on a

subjective view, two people may have the same or widely different CBQ's on the same event, so one person may from time to time preserve or alter his CBQ's on that event in any way to which he is disposed" (1971, p. 49). The really surprising element in the Howson–Urbach proof is the fact that they ignore this gap.

7. The example is spelled out in much more detail elsewhere (Chihara, 1987).

8. Readers familiar with the literature on coherence will, no doubt, see immediately that the foregoing objection will also apply to Dutch Book arguments purporting to show that degrees of beliefs ought to obey the axioms of probability. Of course, there are many other objections to these Dutch Book arguments. See, in this connection, Kennedy and Chihara (1979).

9. I would like to thank Greg Ray and Susan Vineberg for their extensive comments on an earlier version of this chapter. I am confident that the revisions I have made, as a result of their comments, have significantly improved my argumentation.

REFERENCES

Chihara, Charles (1987) "Some Problems for Bayesian Confirmation Theory." *British Journal for the Philosophy of Science* 38: 551–60.

Eells, Ellery (1982) *Rational Decision and Causality*. Cambridge University Press.

Glymour, Clark (1980) *Theory and Evidence*. Princeton University Press.

Howson, Colin, & Urbach, Peter (1989) *Scientific Reasoning: The Bayesian Approach*. Open Court, La Salle, Ill.

Jeffrey, Richard (1970) "Dracula Meets Wolfman: Acceptance vs. Partial Belief." Pp. 157–85 in *Induction, Acceptance, and Rational Belief*, ed. Marshall Swain. D. Reidel, Dordrecht.

Jeffrey, Richard (1983) *The Logic of Decision*, 2nd ed. University of Chicago Press.

Kennedy, R., & Chihara, C. (1979) "The Dutch Book Argument: Its Logical Flaws, Its Subjective Sources." *Philosophical Studies* 36: 19–33.

Levi, Isaac (1970) "Probability and Evidence." Pp. 134–56 in *Induction, Acceptance, and Rational Belief*, ed. Marshall Swain. D. Reidel, Dordrecht.

Mellor, Hugh (1971) *The Matter of Chance*. Cambridge University Press.

Skyrms, Brian (1977) *Choice and Chance*. Wadsworth, Belmont, Calif.

Skyrms, Brian (1987) "Coherence." Pp. 225–41 in *Scientific Inquiry in Philosophical Perspective*, ed. N. Rescher. University Press of America, Lanham, Md.

10

Learning the impossible

VANN MCGEE

Imagine a coin that is going to be tossed infinitely many times. It is by no means a requirement of rationality that we regard the outcomes of the different tosses as independent; we may well regard the outcomes of the early tosses as inductively justifying conclusions about the outcomes of later tosses. On the other hand, it would be bizarre to suppose that rationality forbids us to regard the outcomes as independent; whether or not it is appropriate to do so depends upon what our background theory is and what collateral information we possess. If we do regard the outcomes as independent, then unless the probability of heads on the first toss is either exactly 0 or exactly 1, every particular infinite sequence of heads and tails will have probability 0. But it is certain that some infinite sequence of heads and tails is going to occur; so it is certain that some probability-zero event is going to happen.

This example illustrates the distinction between propositions that have a subjective probability of 0 and propositions that are truly epistemically impossible. If a proposition is utterly impossible epistemically, we shall assign it the probability 0, but it is by no means the case that any proposition to which we assign the probability 0 will be epistemically impossible.

If a proposition we now assign the probability 0 is discovered to be true, we shall revise our system of beliefs. In seeing how we ought rationally to revise our beliefs, standard Bayesian theory is no help. The standard account tells us that where P is our current probability assignment, our new probability of a after we learn b with certainty (without learning anything else) should be $P(a \wedge b)/P(b)$, *provided* $P(b) \neq 0$. The theory gives us no guidance in revising our beliefs when our new information has probability 0.

Two techniques for extending the standard Bayesian theory to solve

I am grateful to Ernest Adams for a number of valuable suggestions.

179

this problem have emerged in the literature. One approach, developed by Brian Skyrms (1980) and David Lewis (1980), is to utilize a nonstandard probability assignment in which those epistemically possible propositions that would ordinarily be assigned 0 probability are instead assigned infinitesimal probabilities; in this way we ensure that whenever b is epistemically possible, $P(a \wedge b)/P(b)$ will be defined. The other approach, developed by Karl Popper (1959, new appendices *ii through *v), is more direct. Rather than define conditional probabilities in terms of absolute probabilities,

$$P(a, b) =_{\mathrm{Df}} \frac{P(a \wedge b)}{P(b)}, \qquad \text{provided } P(b) \neq 0$$

we take the notion of conditional probability as primitive, adopting appropriate axioms, so that the conditional probability of a, given b, can be meaningfully defined even when b has probability 0. We shall see that these two approaches come to the same thing.

We can apply Popper's theory of conditional probabilities to the theory of conditionals. Ernest Adams (1965, 1966, 1975) has proposed that the probability of an English indicative conditional $b \Rightarrow a$ be given by

$$P(b \Rightarrow a) = P(a, b), \qquad \text{provided } P(b) \neq 0$$

Popper's theory of conditional probabilities enables us to drop the proviso that the antecedent must have non-zero probability. Having done so, we get an alternative characterization of the probabilistically valid inferences. An inference is probabilistically valid in the sense described by Adams if and only if any probability assignment that gives the premises probability 1 will also give the conclusion probability 1. This observation will enable us to smooth out some jagged edges that the probabilistic theory of conditionals otherwise encounters at the place where likelihood gives way to certainty. Moreover, it enables us to discern an intimate connection between the probabilistic pragmatics of conditionals and their possible-world semantics.

1. POPPER FUNCTIONS AND NONSTANDARD ANALYSIS

A *Popper function* on a language $\mathscr{L}$ for the classical sentential calculus is a function $C: \mathscr{L} \times \mathscr{L} \to \mathbb{R}$ that meets the following conditions:[1]

(1) For any a and b, there exist c and d with $C(a, b) \neq C(c, d)$.
(2) If $C(a, c) = C(b, c)$ for every c, then $C(d, a) = C(d, b)$ for every d.
(3) $C(a, a) = C(b, b)$.
(4) $C(a \wedge b, c) \leqslant C(a, c)$.

(5) $C(a \wedge b, c) = C(a, b \wedge c) \cdot C(b, c)$.

(6) $C(a, b) + C(\sim a, b) = C(b, b)$, unless $C(b, b) = C(c, b)$ for every c.

Popper makes the further assumption that $\mathscr{L}$ is countable, but this assumption plays no role in the theory, so we drop it.

Among the further properties of Popper functions that Popper derives, we have the following:

(7) Either the function C_b given by $C_b(a) = C(a, b)$ satisfies the standard laws of probability, or else it has constant value 1.

(8) If c implies b and C_c takes the constant value 1, C_b takes the constant value 1.

(9) If a and b are tautologically equivalent and c and d are tautologically equivalent, then $C(a, c) = C(b, d)$.

(10) If $C(c, b) = C(b, c) = 1$, then $C_c = C_b$.

(11) $C_{b \wedge c} = C_{c \wedge b}$.

We get an equivalent system of axioms if axiom (2) is replaced by (9), by (10), or by (11).

C_b is to take the constant value 1 only if b is epistemically utterly impossible. There is no need for even an ideally rational agent to formulate a plan for how she would revise her beliefs upon learning that b, if she is absolutely certain that b is false; in this extreme case, we set $C_b(a)$ equal to 1 arbitrarily.

Having taken conditional probabilities as primitive, we can treat absolute probabilities as defined, defining the absolute probability of a to be $C(a, \mathsf{T})$.[2] Axiom (5) tells us that the conditional probability $C(a, b)$ will be equal to the quotient of the absolute probability of $a \wedge b$ divided by the absolute probability of b, whenever this quotient is defined. Thus Popper's axioms constitute a true generalization of ordinary probability theory.

The alternative approach utilizes a probability function P that, instead of taking its values from the ordinary real number system $\mathbb{R}$, takes its values from the real numbers of a nonstandard model $\mathscr{R}$ of elementary analysis.[3] We write $r \approx s$ to indicate that the difference between r and s is infinitesimal, and if r is in the unit interval of $\mathscr{R}$, we write $\mathrm{st}(r)$ to denote the unique standard real number s such that $r \approx s$. The idea is that if b is an epistemically possible proposition to which we would normally assign the probability 0, we instead assign b an infinitesimal but non-zero probability, so that the quotient $P(a \wedge b)/P(b)$ will be defined. $P(b)$ will be equal to 0 if and only if b is epistemically absolutely impossible.

We shall now see that these two approaches amount to the same thing.

Theorem 1. If P is a nonstandard-valued probability assignment on $\mathscr{L}$

181

then the function $C: \mathscr{L} \times \mathscr{L} \to \mathbb{R}$ given by

(A) $$C(a, b) = \mathrm{st}\left(\frac{P(a \wedge b)}{P(b)}\right), \qquad \text{provided } P(b) \neq 0$$

$$= 1, \qquad \text{otherwise}$$

is a Popper function. Conversely, if C is a Popper function, there is a nonstandard-valued probability assignment P such that

(B) $$P(b) = 0 \quad \text{iff } C_b \text{ is the constant function 1}$$

and

(C) $$C(c, b) = \mathrm{st}\left(\frac{P(c \wedge b)}{P(b)}\right) \quad \text{whenever } P(b) \neq 0$$

Proof. Given a nonstandard probability assignment P, it is effortless to verify that the function $C: \mathscr{L} \times \mathscr{L} \to \mathbb{R}$ defined by (A) satisfies axioms (1)–(6). Our job is to prove the converse: given a Popper function C, to find a nonstandard probability assignment P satisfying (B) and (C).

Suppose that for each sublanguage $\mathscr{L}_F$ of $\mathscr{L}$ that is generated from a finite system of atomic sentences, the restriction of C to $\mathscr{L}_F \times \mathscr{L}_F$ gives rise to a nonstandard probability assignment P_F on $\mathscr{L}_F$ satisfying (B) and (C). Then, by arbitrarily extending the P_F's to $\mathscr{L}$ and then taking an ultraproduct,[4] we get a nonstandard probability assignment on $\mathscr{L}$ that satisfies (B) and (C). Thus there will be no loss of generality if we assume that $\mathscr{L}$ is finitely generated.

Define a sequence $a_0, a_1, \ldots, a_n$ of sentences of $\mathscr{L}$, as follows:

$$a_0 = \top.$$

Given a_k, if C_{a_k} is the constant function 1, set $n = k$, and you are done. Otherwise, let I_k equal the set of sentences b in disjunctive normal form such that b logically entails a_k and $C(b, a_k) = 0$, and let a_{k+1} be the disjunction of the members of I_k.

Lemma. If $j \leqslant n$ and $C(b, a_i) = 0$ for each $i < j$, then b logically entails a_j.

Proof. We prove this by induction on j. The case $j = 0$ is obvious. Suppose, as inductive hypothesis, that $j + 1 \leqslant n$, that $C(b, a_i) = 0$ whenever $i < j + 1$, and that b entails a_j. Then the disjunctive normal equivalent of b is a member of I_j, and so b entails a_{j+1}. ∎

Proof of Theorem 1 (continued). Let $\mathscr{R}$ be a nonstandard model of elementary analysis, and let "Oscar" be an infinitesimal positive number in $\mathscr{R}$.

182

Define a function P from $\mathscr{L}$ to the unit interval of $\mathscr{R}$ by

$$P(b) = \frac{1 - \text{Oscar}}{1 - \text{Oscar}^n} \cdot \sum_{j=0}^{n-1} C(b, a_j) \cdot \text{Oscar}^j$$

It is clear that the function P is a nonstandard probability function. We need to see that it satisfies conditions (B) and (C).

If $P(b) = 0$, then for each $j < n$, $C(b, a_j) = 0$. It follows by the foregoing lemma that b entails a_n. Because, by construction, C_{a_n} is the constant function 1, it follows by (8) that C_b is the constant function 1. This gives the left-to-right direction of (B).

To get the right-to-left direction of (B), suppose that C_b is the constant function 1, and take $j < n$. Because $C(\perp, a_j) = 0$ (because $j < n$) and $C_{b \wedge a_j}$ is the constant function 1 [by (8)], we have

$$0 = C(\perp, a_j) = C(\perp \wedge b, a_j) = C(\perp, b \wedge a_j) \cdot C(b, a_j) = C(b, a_j)$$

Hence, $P(b) = 0$.

Now we need to establish (C). Suppose that $P(b) \neq 0$. Let $j < n$ be the smallest number such that $C(b, a_j) \neq 0$. Then $P(c \wedge b)/P(b)$ is equal to this quotient:

$$\frac{\displaystyle\sum_{i=j}^{n-1} C(c \wedge b, a_i) \cdot \text{Oscar}^{(i-j)}}{\displaystyle\sum_{i=j}^{n-1} C(b, a_i) \cdot \text{Oscar}^{(i-j)}}$$

Here the numerator differs infinitesimally from $C(c \wedge b, a_j)$, the denominator differs infinitesimally from $C(b, a_j)$, and the denominator is not infinitesimal. Hence the quotient differs infinitesimally from $C(c \wedge b, a_j)/C(b, a_j)$; so it is enough to show that $C(c \wedge b, a_j)/C(b, a_j)$ is equal to $C(c, b)$.

We know by the lemma that b entails a_j. Hence, by (5),

$$C(c \wedge b, a_j) = C(c, b \wedge a_j) \cdot C(b, a_j) = C(c, b) \cdot C(b, a_j) \quad \blacksquare$$

In some contexts, it is appropriate to impose the so-called *regularity* requirement, according to which an ideally rational agent will regard b as epistemically impossible only if b is truth-functionally inconsistent. Whether or not it is appropriate to do so depends primarily on whether or not the logical structure of $\mathscr{L}$ is fully captured by its Boolean structure. If we do require that b be epistemically impossible only if b is truth-functionally inconsistent, we shall insist that our nonstandard probability assignment meet the conditon:

$P(b) = 0$ only if b is truth-functionally inconsistent.

And we shall insist that our Popper functions meet the condition

C_b is the constant function 1 only if b is truth-functionally inconsistent.

With these added requirements, Theorem 1 continues to hold.

For some purposes, it is useful to look at sentential calculi in which we can form infinite conjunctions and disjunctions and at probability measures that are countably additive, rather than merely finitely additive. Thus, we may want to add to the definition of a Popper function this further requirement:

(12) If b is the disjunction of $a_0, a_1, a_2, \ldots$ and if $a_i \wedge a_j$ is inconsistent whenever $i \neq j$, then $C(b, c) = \sum_{i=0}^{\infty} C(a_i, c)$.

For the corresponding condition in terms of the nonstandard probability assignments, the first thing that comes to mind is nonsensical:

If b is the disjunction of $a_0, a_1, a_2, \ldots$ and if $a_i \wedge a_j$ is inconsistent whenever $i \neq j$, then $P(b) = \sum_{i=0}^{\infty} P(a_i)$.

The nonstandard model of analysis will not be topologically complete, and so this infinite sum will not normally exist.[5] Instead, the condition we require is this:

If b is the disjunction of $a_0, a_1, a_2, \ldots$ and if $a_i \wedge a_j$ is inconsistent whenever $i \neq j$, then $P(b) \approx \sum_{i=0}^{\infty} \text{st}(P(a_i))$.

It follows immediately from Theorem 1 that with these additional requirements, the two approaches will again coincide.

2. A DUTCH BOOK ARGUMENT

We now know that both prominent approaches give the same answer. Is there any reason to suppose that they give the right answer? As usual in discussing personal probabilities, theory leaps ahead of justification, but a Dutch book argument will give us a little reassurance that we are on the right track. Paul Teller (1973, pp. 222–5) supports the ordinary Bayesian rule for belief revision with a dynamic version of the Dutch book argument that he attributes to Lewis, and we can adapt this argument to our purposes here.

Let the finite set of sentences $\{b_j : j \in J\}$ constitute a complete description of every possible course of experience the agent might have between now and time t. "Possible" here refers to epistemic possibility within the system

184

of beliefs of an ideally rational agent; the agent regards it as an utter certainty that one of the b_j's will occur. The b_j's are assumed to be mutually incompatible.

Let $\mathscr{L}$ be a finitely generated language that includes each of the B_j's and that consists entirely of sentences whose truth values will eventually be discovered, so that a bet on a sentence of $\mathscr{L}$ will always be settled. Let $\{a_i: i \in I\}$ be the set of state descriptions[6] for $\mathscr{L}$, and let $\{a_i: i \in I'\}$ consist of those state descriptions for $\mathscr{L}$ that the agent regards as epistemically possible. If we assume that the agent's belief system is regular, we shall have $I' = I$.

For c a sentence of $\mathscr{L}$, let $BQ(c)$ be the number r such that the agent is indifferent between giving or receiving \$$r$ in exchange for an arrangement in which one gets \$1 if c is true, but nothing if c is false. If we treat monetary gain as proportional to gain in utility, we can say that a situation in which the agent pays \$$p$, in exchange for which she receives \$$q$ if c is true, but nothing if c is false, will be *advantageous* for the agent if and only if (iff) $p < q \cdot BQ(a)$. Similarly, an arrangement in which the agent receives \$$p$ in exchange for agreeing to pay \$$q$ if a is true, but nothing if a is false, will be advantageous iff $p > q \cdot BQ(a)$. We can consolidate our treatment of the two kinds of arrangements by describing the latter arrangement as one in which the agent has paid \$$-p$ in exchange for receiving \$$-q$ if a is true, but nothing if a is false.

All the "bets" we shall discuss will be arrangements of this sort, entered into either now or at time t. Let $BQ_{b_j}(c)$ be the number r such that if b_j accurately describes the agent's experiences between now and time t, then at time t the agent will be indifferent between giving or receiving \$$r$ in exchange for an arrangement in which one receives \$1 if c is true, but nothing if c is false.

A *Dutch book* is a finite system of bets with the property that even though the agent regards each of the bets individually as advantageous, there is some possible (as judged by the agent) situation in which the agent will suffer a net loss from the whole system of bets, and there is no possible situation in which the agent will enjoy a net gain.[7] The thesis underlying the Dutch book arguments is that if the agent makes herself vulnerable to a Dutch book, so that a cunning bookie (knowing only the agent's dispositions to betting behavior and the rule by which the agent will change her betting behavior when she acquires new evidence, and not knowing anything at all about the propositions that are the subjects of the bets) can entice the agent into a situation in which she might possibly lose and she cannot possibly win, then there is something drastically wrong with the way the agent places her bets. Either there is some time at which she bets incoherently, or else her rule for changing her betting behavior when she gets new evidence is badly defective.

Theorem 2. The agent will be immune to a Dutch book if and only if there is a Popper function C with $C(c, \top) = BQ(c)$, with $C(c, b_j) = BQ_{b_j}(c)$, and with $C(\bot, a_i) = 0$ iff $i \in I'$.

Proof. The static Dutch book theorem[8] tells us that the agent will be vulnerable to a Dutch book unless the following conditions are met:

(a) BQ obeys the standard laws of probability, and $\sum_{j \in J} BQ(b_j) = 1$.
(b) BQ_{b_j} obeys the standard laws of probability, and $BQ_{b_j}(b_j) = 1$.

Similar arguments straightforwardly show that these conditions are also needed:

(c) $BQ(c \wedge b_j) = BQ_{b_j}(c) \cdot BQ(b_j)$.
(d) If $i \in I \sim I'$, $BQ_{b_j}(a_i) = 0$.

We know that if the agent is immune to a Dutch book, then conditions (a)–(d) will be met. We need to show that if conditions (a)–(d) are met, then there is a Popper function C with $C(c, \top) = BQ(c)$, with $C(c, b_j) = BQ_{b_j}(c)$, and with $C(\bot, a_i) = 0$ iff $i \in I'$.

Let $\mathscr{R}$ be a nonstandard model of analysis, and let "Oscar" be an infinitesimal of $\mathscr{R}$. Define a function $\tilde{P}$ assigning a nonstandard real number to each state description, as follows:

$$\begin{aligned}
\tilde{P}(a_i) &= BQ(a_i) \quad \text{if } BQ(a_i) \neq 0 \\
&= BQ_{b_j}(a_i) \cdot \text{Oscar} \quad \text{if } BQ_{b_j}(a_i) \neq 0 \text{ but } BQ(a_i) = 0 \\
&= \text{Oscar}^2 \quad \text{if } i \in I' \text{ but } BQ_{b_j}(a_i) = 0 \text{ for each } j \in J \\
&= 0 \quad \text{if } i \notin I'
\end{aligned}$$

Let

$$r = 1 + \text{Oscar} \cdot (\text{the number } j\text{'s in } J \text{ with } BQ(b_j) = 0)$$
$$+ \text{Oscar}^2 \cdot (\text{the number of } i\text{'s in } I' \text{ with } BQ_{b_j}(a_i) = 0 \text{ for each } j \in J)$$

Now define, for c in $\mathscr{L}$,

$$P(c) = \frac{1}{r} \sum \{\tilde{P}(a_i): a_i \text{ implies } c\}$$

It is routine to verify that P is a nonstandard probability assignment under which $P(c) \approx BQ(c)$, under which $P(c \wedge b_j)/P(b_j) \approx BQ_{b_j}(c)$ whenever $P(b_j) \neq 0$, and under which $P(a_i) \neq 0$ iff $i \in I'$. It follows by Theorem 1 that

186

setting

$$C(c,d) = \text{st}\left(\frac{P(c \wedge d)}{P(d)}\right) \quad \text{if } P(d) \neq 0$$

$$= 1 \quad \text{if } P(d) = 0$$

will give us the Popper function we desire.

For the other direction, suppose that there is a Popper function C with $C(c, \mathsf{T}) = \text{BQ}(c)$, with $C(c, b_j) = \text{BQ}_{b_j}(c)$, and with $C(\bot, a_i) = 0$ iff $i \in I'$. We know by theorem 1 that there is a nonstandard probability assignment P with $P(d) = 0$ iff C_d takes the constant value 1 and with $P(c \wedge d)/P(d) \approx C(c,d)$ whenever $P(d) \neq 0$.

We want to show that the agent's present and future dispositions to betting behavior will not permit her to fall prey to a Dutch book. That is, we want to show that it will not be possible for a bookie who knows the agent's current dispositions to betting behavior, and who also knows how the agent's dispositions to betting behavior will change as she acquires new evidence between now and time t, to entice the agent into a Dutch book. A disposition to change her beliefs in such a way that if b_j occurs, then at t the agent will be willing to pay p in exchange for getting q if c is true, but nothing if c is false, is equivalent, as far as determining the outcomes of bets is concerned, to a current disposition to pay p to participate in a conditional bet that will pay her q if $(c \wedge b_j)$ is true, that will pay her nothing if $(\sim c \wedge b_j)$ is true, and that will be called off (with her payment refunded) if b_j is false. If K is such a conditional bet, we may define the *net payoff function* for K, f_K, assigning a real number to each state description of $\mathcal{L}$, by

$$f_K(a_i) = (q - p) \quad \text{if } a_i \text{ implies } (c \wedge b_j)$$

$$= -p \quad \text{if } a_i \text{ implies } (\sim c \wedge b_j)$$

$$= 0 \quad \text{if } a_i \text{ implies } \sim b_j$$

The agent will regard such a conditional bet as advantageous just in case the sum

$$\sum \{f_K(a_i) \cdot C(a_i, b_j): a_i \text{ implies } b_j\}$$

is positive.

If L is a simple bet in which the agent now pays p in exchange for getting q if c is true, but nothing if c is false, the net payoff function f_L will be given by

$$f_L(a_i) = (q - p) \quad \text{if } a_i \text{ implies } c$$

$$= -p \quad \text{if } a_i \text{ implies } \sim c$$

The agent will regard such a simple bet as advantageous just in case the sum

$$\sum \{f_L(a_i) \cdot C(a_i, \mathsf{T}) : i \in I\}$$

is positive.

In either case, if K is either a simple bet or a conditional bet that the agent finds advantageous, we must have

$$\sum_{i \in I} f_K(a_i) \cdot P(a_i) > 0$$

Thus, if $\mathcal{K}$ is a finite system of simple or conditional bets, each of which the agent finds advantageous, we must have

$$0 < \sum_{K \in \mathcal{K}} \sum_{i \in I} f_K(a_i) \cdot P(a_i) = \sum_{i \in I} \left(\sum_{K \in \mathcal{K}} f_K(a_i) \right) \cdot P(a_i)$$

$$= \sum_{i \in I'} \left(\sum_{K \in \mathcal{K}} f_K(a_i) \right) \cdot P(a_i)$$

So there is at least one $i_0 \in I'$ for which $\sum_{K \in \mathcal{K}} f_K(a_{i_0})$ is positive. So there is at least one possible situation in which the agent will come out of the system of bets a winner. ∎

Just as the original Lewis–Teller argument gives us reason to suppose that the standard Bayesian rule correctly tells us how to revise our beliefs when we acquire new evidence that has non-zero probability, the adapted version of the argument gives us reason to think that Popper has given us the right generalization of the rule to use when our new evidence has prior probability 0.

We shall now look at an application of Popper functions to the logic of conditionals.

3. THE LOGIC OF LIKELIHOOD AND THE LOGIC OF CERTAINTY

In proposing his probabilistic account, Adams challenged a deeply entrenched orthodoxy, according to which, though the logic of subjunctive conditionals was very much in question, the logic of indicative conditionals was simple: An indicative conditional is true if and only if its antecedent is false or its consequent true, and a pattern of inference is valid if and only if no instance of the pattern could ever have true premises and a false conclusion. This criterion sanctions patterns of reasoning that are utterly repugnant to English-speaking reasoners, so Adams offered a new logic of conditionals that much more closely reflects the inferences that speakers of English actually employ.

188

Adams proposed that a pattern of inference be considered valid iff it is not possible for its premises to be highly likely without its conclusion also being highly likely. Specifically, he proposed the following:

Logic of likelihood criterion[9]

An inference is *probabilistically valid* if and only if, for any positive ε, there exists a positive δ such that, under any probability assignment under which each of the premises will have probability greater than $1 - \delta$, the conclusion will have probability at least $1 - \varepsilon$.

To apply this standard to give a logic of conditionals, Adams needed to say what the probability of a conditional is:

Original Adams hypothesis[10]

The probability of a conditional $a \Rightarrow b$ will be $P(a \wedge b)/P(b)$ if $P(b)$ is nonzero, and it will be equal to 1 if $P(b)$ equals 0.

This rule gives us the probability of so-called *simple* conditionals, that is, conditionals whose components are what Adams calls *factual* or conditional-free. How to extend this rule to compounds of conditionals has proved to be quite a difficult problem, and we shall follow Adams in setting it aside, restricting our attention to inferences in which a finite system of premises that are either factual statements or simple conditionals leads to a conclusion that is either a factual statement or a simple conditional.

The logical system that Adams develops describes, with remarkable fidelity, the modes of inference that ordinary speakers employ. When we take the limit as the uncertainties go to 0, however, the harmony is abruptly broken.

Logic of certainty criterion[11]

An inference is *strictly valid* if and only if its conclusion has probability 1 under any probability assignment under which its premises each have probability 1.

The strictly valid inferences are not those described by Adams' theory, but those described by the orthodox theory, which treats the English conditional as the material conditional.

This raises an ugly suspicion. The failures of the classically valid modes of inference appear only when we are reasoning from premises that are less than certain (in the sense of having probability less than 1) to a conclusion that is also less than certain. Once we become certain of our premises, we can deduce the classically sanctioned consequences with

189

assurance. This suggests that perhaps the classical inferences are really valid after all, and the appearance of invalidity is merely a side effect of the uncertainty of our premises. So long as we are in uncertain circumstances, we are unable to distinguish inferences that are truly invalid from valid inferences that we sometimes resist as a way of coping with uncertainty. Only when uncertainty disappears are we able properly to distinguish valid from invalid inferences, and when we do so, the correctness of the classical account becomes apparent.

In determining that the strictly valid inferences are the classical ones, what is important is not Adams' central thesis that the probability of a conditional is the conditional probability of the consequent, given the antecedent, whenever the latter is defined. Rather, it is the default condition that assigns the conditional the probability 1 when the conditional probability is undefined. This default condition does not reflect English usage, nor was it intended to do so. English speakers are not inclined to say the following:

> If the sequence of tosses forever alternates heads, tails, heads, tails, and so on, then the coin has two heads.

Nor do they assent to this:

> If the pointer lands exactly at the four o'clock position, then my father-in-law is the Archangel Gabriel.

On the contrary, the default condition, as Adams notes (1975, p. 40), is merely "arbitrarily stipulated" as a way of setting aside a special case that is far removed from the central focus of concern. Yet the default condition has caused a good deal of mischief; so it is time to look for an alternative.

We have the alternative ready to hand. Popper functions offer a natural generalization of the ordinary notion of conditional probability in terms of which the singularities that otherwise appear at the edge of certainty no longer appear.

Improved Adams hypothesis

The probability of a simple conditional $a \Rightarrow b$ is $C(b, a)$, where C is a Popper function on the set of factual sentences.

Theorem 3. In terms of the improved Adams hypothesis, the following are equivalent, for any inference from a finite set of factual or simple conditional premises to a factual or simple conditional conclusion:

(i) The inference is probabilistically valid.
(ii) The inference is strictly valid.

190

(iii) There is no Popper function that assigns the premises probability 1, and yet assigns the conclusion probability 0.

Proof. The theorem follows directly from the results of Adams (1966), together with Theorem 1. Because we can replace any factual statement f that occurs in the argument by the conditional $\top \Rightarrow f$, we can assume that the argument consists entirely of conditionals, so that it has the form

$$c_1 \Rightarrow d_1$$
$$c_2 \Rightarrow d_2$$
$$\cdots\cdots\cdots$$
$$c_n \Rightarrow d_n$$

$$\therefore a \Rightarrow b$$

Proposition (i) implies proposition (ii). Because the inference is probabilistically valid, for each positive integer k there is a positive number δ_k such that any Popper function that assigns each of the premises a probability greater than $1 - \delta_k$ is sure to assign the conclusion a probability greater than $1 - 1/k$.

Let C be a Popper function that assigns the value 1 to each of the premises. Let P be a nonstandard probability function, with $P(e) = 0$ iff $C(\bot, e) = 1$ and with $C(f, e) = \mathrm{st}(P(f \wedge e)/P(e))$ whenever $P(e) \neq 0$. For each k and i, $1 - P(d_i \wedge c_i)/P(c_i)$ is infinitesimal, so it is certainly less than δ_k. Hence $1 - P(b \wedge a)/P(a)$ is less than $1/k$. Because k was arbitrary, $1 - P(b \wedge a)/P(a)$ must be infinitesimal, and $C(b, a)$ must be equal to 1.

Proportion (ii) implies proposition (iii). Immediate.

Proposition (iii) implies proposition (i). Adams (1966, p. 308) shows that if the inference is not valid, then for any positive ε there is a probability assignment that assigns each of the premises a probability greater than $1 - \varepsilon$ and that assigns the conclusion a probability less than ε. Taking ε to be infinitesimal, this gives us a nonstandard probability assignment in which the value assigned to each of the premises differs infinitesimally from 1, and the value assigned to the conclusion differs infinitesimally from 0. This induces a Popper function according to which the premises have probability 1, and the conclusion probability 0. ∎

The original Adams hypothesis is, in effect, the improved Adams hypothesis together with this added assumption:

$$C(\bot, e) = 1 \quad \text{whenever } C(e, \top) = 0$$

This extra assumption is entirely unjustified, and when we drop it, the disparity between probabilistic validity and strict validity disappears.[12]

We now have before us three concepts of logical validity: probabilistic validity, strict validity, and the traditional conception:

Logic of truth criterion

An inference is *semantically valid* if and only if its logical form ensures that its premises cannot all be true without its conclusion also being true.

Of the three conceptions of validity, the semantic conception has figured most prominently in philosophers' reasoning about logic and language, but it is the most difficult criterion to apply in evaluating particular arguments, for to apply it we already need to know the conditions under which a sentence of a language is true. For conditionals, it is very much in question what their truth conditions are, or even whether they have truth values, and so it is very much in question which inferences involving conditionals are truth-preserving.

In trying to decide which statements are, within a given speaker's system of beliefs, highly probable, we have a lot of evidence to go on. We see which statements a speaker asserts and which she denies, which remarks by others she approves and which she scorns, which statements she continues to uphold in the face of criticism and which she modifies or withdraws, and how she changes what she is willing to assert or deny in the light of new evidence. This evidence does not give us anything close to the exquisitely finely calibrated judgments of likelihood postulated by subjective probability theory, but it does enable us crudely to distinguish statements that are highly probable for the speaker from those that are highly improbable or indifferent, and so it enables us to recognize arguments that lead from highly probable premises to highly probable conclusions. We still have a great distance to go to get from arguments that do, in fact, lead, within a given speaker's system of beliefs, from highly probable premises to highly probable conclusions, to arguments that ought, in principle, to lead from highly probable premises to highly probable conclusions – but again we have evidence to guide us. We can distinguish cases in which the speaker explicitly refuses to accept a conclusion from those in which she never even thinks about it; we can distinguish those inferences that are products of careful and sober deliberation; we can see which inferences hold up well under criticism; we can determine which inferences are approved by the community at large; and we can see which inferences the speaker herself approves, looking back on them, years later. These observations give us something to start from in constructing the logic of likelihood.

192

Linguistic behavior gives us direct evidence as to what propositions are, for a given speaker, highly probable; this evidence gives us a starting point for the logic of likelihood. We do not have similar evidence from linguistic behavior as to what propositions are true, and so we do not have a similar starting point for the logic of truth. We have direct evidence as to what propositions the speakers of the language *hold* true, but that is just another way of saying that we have direct evidence for what propositions are, for the speakers of the language, highly probable. Truth is very much less tangible than subjective likelihood, and so our procedure in constructing the logic of truth will be much less direct.

The theory of semantic validity and the theory of probabilistic validity seek to describe and explain precisely the same body of linguistic behavioral data. Indeed, there is good reason to think that, assuming that our understanding of the notion of truth is governed by Tarski's schema[13]

(T) $\qquad\qquad$ $\ulcorner p \urcorner$ is true if and only if p

the two theories will sanction precisely the same inferences, so that an argument is probabilistically valid just in case it is not possible for it to have true premises and a false conclusion.

A reason for anticipating that semantically valid arguments will also be probabilistically valid is the observation that the premises of an argument, together with the relevant instances of schema (T) and the thesis that the argument does not have true premises and a false conclusion, probabilistically entail the conclusion of the argument; so, if the argument is semantically valid but not probabilistically valid, the blame must lie either with some uncertainty as to the argument's semantic validity or with some misgivings about schema (T).

For the converse, we need Adams' (1975, p. 51) notion of *probabilistic consistency*: A finite set of sentences is probabilistically consistent iff, for any positive ε, there is a probability assignment that assigns to every member of the set a probability greater than $1 - ε$. If an argument is probabilistically valid, then the relevant instances of schema (T) are probabilistically inconsistent with the thesis that the argument has true premises and a false conclusion.

These considerations are entirely general, having nothing in particular to do with conditionals. They would lead us to expect probabilistic validity and semantic validity to coincide. This expectation is severely disappointed by the orthodox theory, and, indeed, one of Adams' principal dissatisfactions with the orthodox theory arises from the many striking examples of classically semantically valid inferences that would be utterly repugnant to English speakers. We would expect the logic of likelihood and the logic of truth to coincide.

193

Alternatives to the orthodox semantics have been proposed, the most prominent of which are the possible-world accounts, invented by Robert Stalnaker (1968) and further developed by Lewis (1973) and others. According to Stalnaker, a conditional is true in a possible world w iff the consequent is true in the possible world most similar to w in which the antecedent is true; the conditional will also count as true in w, by default, if there is no world at all similar to w in which the antecedent is true. He makes this precise by postulating a *selection function* that picks out, for each world w and sentence p, the world most similar to w in which p is true; if there is no world at all similar to w in which p is true, the selection function picks out the *impossible world*, in which everything is true. Lewis' theory differs from Stalnaker's in permitting ties in the similarity relation, so that a conditional is true at w iff the consequent is true in the *worlds* most similar to w in which the antecedent is true.

Adams (1977) has studied the relation between his own system and those of Stalnaker and Lewis, and he has found that they have precisely the relation we want: An inference leading from a finite set of factual or simple conditional premises to a factual or simple conditional conclusion is probabilistically valid iff it is semantically valid within the Stalnaker semantics iff it is semantically valid within the Lewis system. Thus, if we adopt either the Stalnaker semantics or the Lewis semantics, our three conceptions of logical validity will coalesce.

Attempts to find a closer relation between the probabilistic pragmatics of conditionals and their possible-world semantics have had mixed results. The slogan "probability equals probability of truth" has been taken[14] to refer to the hypothesis that there is a way of assigning a probability to each pair $\langle w, f \rangle$, where w is a possible world and f is a selection function, in such a way that the probability of a sentence p is the sum of the probabilities assigned to all pairs $\langle w, f \rangle$ such that, with f as the selection function, p is true in w. Adams (1975, pp. 5–8)[15] has shown that if we are allowed to distribute the probabilities among the world/selection-function pairs any way we like, the hypothesis will be false. McGee (1989, pp. 531–2) shows that there is a fairly natural way to restrict the distribution of probability among the worlds so that the hypothesis is true. At present, the issue is still very much in doubt.

Indeed, it is much in doubt whether there even ought to be a semantics of conditionals, for Adams himself thinks that conditionals are neither true nor false. This is not an issue we need to resolve here. Quite the contrary, the point I want to emphasize here is that one can understand and employ the logic of likelihood before taking any position whatever with regard to the truth conditions for conditionals. This is why even those who are firmly convinced that an adequate logical understanding is not possible without a theory of truth ought nonetheless to regard Adams' work as so very valuable. The logic of likelihood is a self-contained pragmatics. You can develop the logic of likelihood before you start work constructing the theory of truth, and completing the former project will provide invaluable guidance in carrying out the latter.

5. THE LOGIC OF CERTAINTY AND THE LOGIC OF TRUTH

Dual to the three notions of validity we have discussed, there are three notions of consistency. We have already mentioned *probabilistic consistency*: A set of sentences is probabilistically consistent if their joint probability can be made to be arbitrarily close to 1. We can define a set of sentences to be *strictly consistent* iff there is a Popper function that assigns probability 1 to all the members of the set. Finally, a set of sentences is *semantically consistent* iff it is possible for all its members to be true.

It follows from Adams (1977) that a finite set of factual and simple conditional sentences is probabilistically consistent iff it is semantically consistent according to either the Stalnaker system or the Lewis system, and it follows from Theorem 3 that a finite set of factual and simple conditional sentences is probabilistically consistent iff it is strictly consistent. Thus, we see that any finite set of factual and simple conditional sentences will be strictly consistent iff it is consistent in either the Stalnaker system or the Lewis system. Because strict consistency is compact,[16] as are both systems of possible-world semantics,[17] we conclude that an arbitrary set of factual and simple conditional sentences will be strictly consistent iff it is consistent in either the Stalnaker system or the Lewis system. Thus it is coherent to regard every member of a given set of factual and simple conditional sentences as a certainty iff it is consistent to regard every member of the set as true.

This observation gives a partial vindication of the slogan "certainty equals certainty of truth," but a sharper result is possible:

Definition. A Popper function is *decisive* iff it takes only the values 0 and 1.

Theorem 4. Given a set Γ of factual and simple conditional sentences. There is a decisive Popper function under which all and only the members of Γ are assigned the value 1 if and only if there is a Stalnaker model in which all and only the members of Γ are true.

Proof. Make Γ into a set of simple conditionals by replacing e by $\top \Rightarrow e$. Given a decisive Popper function C under which all and only the members

195

of Γ are assigned the value 1, define a function f as follows:

If C_b is not the constant function 1, let $f(b) = \{a : C(a, b) = 1\}$.
If C_b is the constant function 1, $f(b)$ will be undefined.

Now obtain a Stalnaker model by applying the following:

Lemma. Given a function f assigning a maximal consistent set of factual sentences to each member of a certain nonempty set of factual sentences, satisfying the following three conditions:

If $a \in \mathrm{Dom}(f)$, then $a \in f(a)$
If $a \in \mathrm{Dom}(f)$ and $b \in f(a)$, then $b \in \mathrm{Dom}(f)$.
If a and b are both in $\mathrm{Dom}(f)$ and $a \in f(b)$ and $b \in f(a)$, then $f(a) = f(b)$.

Then there is a Stalnaker model whose selection function g satisfies these conditions:

If $a \in \mathrm{Dom}(f)$, then $g(a, \text{the actual world}) = f(a)$.
If $a \notin \mathrm{Dom}(f)$, then $g(a, \text{the actual world}) = \text{the impossible world}$.

Proof. Take the set of worlds to consist of the maximal consistent sets of factual sentences, together with the impossible world, and say that a factual sentence is true in a maximal consistent set of factual sentences iff it is a member of the set. Let the actual world be $f(\mathsf{T})$. Define the selection function g as follows: If w is a maximal consistent set other than $f(\mathsf{T})$, set

$$g(a, w) = w, \qquad \text{if } a \text{ is true in } w$$
$$= \text{the impossible world}, \qquad \text{if } a \text{ is false in } w$$

Moreover, set

$$g(a, f(\mathsf{T})) = f(\mathsf{T}), \qquad \text{if } a \text{ is true in } f(\mathsf{T})$$
$$= f(\text{the result of replacing each ``}{\Rightarrow}\text{'' in } a \text{ by ``}{\supset}\text{''}),$$
$$\text{if } a \text{ is false in } f(\mathsf{T}) \text{ and the result of replacing}$$
$$\text{each ``}{\Rightarrow}\text{'' in } a \text{ by ``}{\supset}\text{'' is in } \mathrm{Dom}(f)$$
$$= \text{the impossible world}, \qquad \text{otherwise}$$

It is easy to check that this is a Stalnaker model with the desired properties. ∎

Proof of Theorem 4 (continued). For the other direction, given a Stalnaker model in which all and only the members of Γ are true, define a decisive Popper function C by stipulating

$$C(a, b) = 1, \qquad \text{if } b \Rightarrow a \text{ is true in the model}$$
$$= 0, \qquad \text{otherwise.} \quad ∎$$

196

Notice that if we use Lewis models in place of Stalnaker models, the theorem will no longer be true, because condition (6) of the definition of a Popper function will no longer be met.

One favorite way to think about truth is that truth is the ideal limit of inquiry. Truth is what an epistemically ideally situated agent gets in the limit as belief gives way to knowledge and doubt gives way to certainty. This picture, that the truth is what would be believed by a believer whose beliefs were as good as they could possibly get, is precisely what Theorem 4 gives us.

The problem we have been examining, how to revise one's system of beliefs upon obtaining new evidence that had prior probability 0, is not a problem that has any great practical significance. Whereas there are a great many epistemically possible theses that have probability 0, it seldom if ever happens that we learn one of them. To do so, we would have to observe an infinite sequence of coin tosses or note the position of a pointer with infinite precision.

The significance of the problem is entirely theoretical. Solving it enables us to avoid singularities that otherwise occur in the limit as our degrees of uncertainty get closer and closer to 0. In particular, it enables the logic of certainty to serve as a bridge between the logic of likelihood and the logic of truth. Decisive Popper functions are the natural limit, as opinion gives way to certainty, of the logic of likelihood, and for the limited class of sentences we have been looking at, a Stalnaker model can be directly constructed from a decisive Popper function. Thus, the use of Popper functions in place of ordinary conditional probabilities is, in a small way, a theoretical advance, for it connects some loose ends and smooths some rough edges.

NOTES

1. For simplicity, I speak of subjective probabilities as numerical values assigned to sentences, even though it would be more accurate to speak of probability as measuring the degree of belief attached to the proposition expressed by a sentence within a particular context of utterance. By "proposition," incidentally, I refer to the objects of belief, whatever their metaphysical character; in particular, I do not take it for granted that propositions have truth values.
2. $\top$ is a tautology, and $\bot$ is a contradiction.
3. For the construction of a nonstandard model of analysis, see Robinson (1966) or Skyrms (1980, appendix 4).
4. For the ultraproduct construction, see Bell and Slomson (1969).
5. Suppose that the disjunction of the a_i's is tautological and that none of the $P(a_i)$'s is infinitesimal, and let $\sigma(n)$ equal $\sum_{i=0}^{n} P(a_i)$. The increasing sequence $\langle \sigma(0), \sigma(1), \sigma(2), \ldots \rangle$ will be bounded above by $1 - \varepsilon$ for each infinitesimal ε. The infinite sequence cannot have a limit in $\mathfrak{R}$, because if r were its limit, we could define the set of standard natural numbers within $\mathfrak{R}$ as $\{k: 1 - 1/k < r\}$.

197

It might happen that within $\mathfrak{R}$ we can define an increasing function $\tilde{\sigma}$ assigning a member of the unit interval of $\mathfrak{R}$ to each of the natural numbers of $\mathfrak{R}$ in such a way that $\tilde{\sigma}(k) = \sigma(k)$ for each standard k. $\tilde{\sigma}$ will have a limit in $\mathfrak{R}$, but the limit will be of no use, because it will depend crucially upon the values $\tilde{\sigma}$ assigns to the nonstandard integers.

6. A *state description* is a long conjunction that includes, for each atomic sentence of $\mathscr{L}$, either the sentence itself or its negation as a conjunct.

7. If we were to adopt the stricter definition according to which a Dutch book is a system of bets from which the agent is sure to suffer a net loss, no matter what, then to ensure that the agent would be immune to a Dutch book it would be enough to require that BQ satisfy the standard laws of probability and that $BQ_{b_j}(c)$ be equal to $BQ(c \wedge b_j)/BQ(b_j)$ whenever $BQ(b_j)$ is non-zero.

8. See de Finetti (1937) or Skyrms (1975, chap. 6).

9. From Adams (1966, p. 274).

10. From Adams (1966, p. 273). In his 1975 book, Adams avoids making a decision about what to do about probability-zero antecedents by looking only at probability assignments in which every consistent factual sentence has positive probability and by banning from the language conditionals with inconsistent antecedents.

11. From Adams (1966, p. 274). Ramsey (1926) uses the phrases "logic of certainty" and "logic of truth," but his usage is quite different from our usage here.

12. Adams has obtained many further results on the probabilistic logic of conditionals, most of which generalize quite nicely to results either about nonstandard probability functions or about Popper functions. For example, Adams (1983) has an elegant explanation of the fact that many classically valid modes of inference, though not flawlessly valid, turn out, in practice, to give satisfactory results almost all the time. Contraposition, disjunctive syllogism, and simplification of disjunctive antecedents are not above reproach, but it is not easy to find counterexamples. Adams' explanation is that whenever a classically valid argument (with factual and simple conditional premises and a simple conditional conclusion) has highly probable premises without having a highly probable conclusion, the antecedent of the conclusion is highly improbable. Thus, classically valid inferences seldom lead us astray because we seldom employ them in situations in which the antecedent of the conclusion is almost certainly false. In terms of nonstandard models, we can say that whenever the probabilities of the premises of a classically valid argument differ infinitesimally from 1, the probability of the conclusion will also differ infinitesimally from 1, unless the probability of the antecedent of the conclusion is infinitesimal. A Popper function C that assigns the value 1 to each of the premises of a classically valid argument with conclusion $a \Rightarrow b$ will set $C(b, a)$ equal to 1 unless it sets $C(a, \top)$ equal to 0. Further results along the same lines, which generalize without incident to results about nonstandard probability assignments and about Popper functions, can be found in Adams (1986).

13. From Tarski (1944, p. 350). Following Tarski, I take the "if and only if" of schema (T) to indicate material equivalence, thus avoiding having to worry about conditionals that have conditional components. One cannot consistently accept the unrestricted application of schema (T), because of the liar paradox, but one could accept its application to a suitably restricted object language.

14. By Adams (1975, p. 5).

15. The argument is an adaptation of the "triviality theorem" of Lewis (1976).

16. If for every finite subset Δ of the set Γ of sentences there is a nonstandard probability assignment that allots to each of the members of Δ a probability that differs infinitesimally from 1, then by taking an ultraproduct, we get a nonstandard probability assignment that allots to each member of Γ a probability differing infinitesimally from 1.
17. This follows directly from the canonical models construction given by Lewis (1971).

REFERENCES

Adams, Ernest W. (1965) "On the Logic of Conditionals." *Inquiry* 8: 166–97.
Adams, Ernest W. (1966) "Probability and the Logic of Conditionals." Pp. 265–316 in *Aspects of Inductive Logic*, ed. J. Hintikka & P. Suppes. North Holland, Amsterdam.
Adams, Ernest W. (1975) *The Logic of Conditionals*. D. Reidel, Dordrecht.
Adams, Ernest W. (1977) "A Note on Comparing Probabilistic and Modal Logics of Conditionals." *Theoria* 43: 186–94.
Adams, Ernest W. (1983) "Probabilistic Enthymemes." *Journal of Pragmatics* 7: 283–95.
Adams, Ernest W. (1986) "Remarks on the Semantics and Pragmatics of Conditionals." Pp. 169–78 in *On Conditionals*, ed. E. Traugott et al. Cambridge University Press.
Bell, A. M., & Slomson, A. B. (1969) *Models and Ultraproducts*. North Holland, Amsterdam.
de Finetti, Bruno (1937) "La Prévision: Ses Lois Logiques, Ses Sources Subjectives." *Annales de l'Institut Henri Poincaré* 7: 1–68.
Harper, William, Stalnaker, Robert, & Pearce, Glenn (eds.) (1981) *Ifs*. D. Reidel, Dordrecht.
Hintikka, Jaakko, & Suppes, Patrick (eds.) (1966) *Aspects of Inductive Logic*. North Holland, Amsterdam.
Lewis, David K. (1971) "Completeness and Decidability of Three Logics of Counterfactual Conditionals." *Theoria* 37: 74–85.
Lewis, David K. (1973) *Counterfactuals*. Blackwell, Oxford.
Lewis, David K. (1976) "Probabilities of Conditionals and Conditional Probabilities." *Philosophical Review* 85: 297–315; reprinted in Harper et al. (eds.) (1981) *Ifs*. D. Reidel, Dordrecht.
Lewis, David K. (1980) "A Subjectivist's Guide to Objective Chance." Pp. 267–99 in *Ifs*, ed. W. Harper, R. Stalnaker, & G. Pearce. D. Reidel, Dordrecht.
McGee, Vann (1989) "Conditional Probabilities and Compounds of Conditionals." *Philosophical Review* 98: 485–541.
Popper, Karl R. (1959) *The Logic of Scientific Discovery*, 2nd ed. Basic Books, New York.
Ramsey, F. P. (1926) "Truth and Probability." Pp. 156–98 in his *The Foundations of Mathematics and Other Logical Essays* (1931), ed. R. B. Braithwaite. Routledge & Kegan Paul, London. Reprinted (1978, pp. 58–100) in his *Foundations*, ed. D. H. Mellor. Routledge & Kegan Paul, London, and in (1964, pp. 61–92) *Studies in Subjective Probability*, ed. H. E. Kyburg & H. E. Smokler. Wiley, New York.
Robinson, Abraham (1966) *Nonstandard Analysis*. North Holland, Amsterdam.
Skyrms, Brian (1975) *Choice and Chance*, 2nd ed. Wadsworth, Belmont, Calif.
Skyrms, Brian (1980) *Causal Necessity*. Yale University Press, New Haven, Conn.
Stalnaker, R. (1968) "A Theory of Conditionals." Pp. 98–112 in *Studies in Logical Theory*, ed. N. Rescher. Blackwell, Oxford.
Tarski, Alfred (1944) "The Semantic Conception of Truth." *Philosophy and Phenomenological Research* 4: 341–76.
Teller, Paul (1973) "Conditionalization and Observation." *Synthese* 26: 218–58.

11

A brief survey of Adams' contributions to philosophy

PATRICK SUPPES

Before making comments on Ernest Adams' theory of conditionals, I am going to exercise a prerogative reserved at least for his thesis adviser – Ernie was my first Ph.D. student – to give a survey of Adams' contributions to a wide range of topics in philosophy.

I have divided his work under five headings. The divisions may be somewhat arbitrary, but still they show well the range of his interests. The five divisions are as follows: the foundations of physics; utility theory and game theory; general measurement theory; the foundations of geometry; and language and logic, especially conditionals.

1. FOUNDATIONS OF PHYSICS

Adams made a significant contribution to axiomatic work in the foundations of physics with his thesis, giving a representation theorem for rigid bodies in terms of finite systems of particles. In spite of the naturalness of the basic result, it had not previously been proved in the literature in general form. He also began there an important aspect of his thinking, namely, a series of critical reflections on the problem of characterizing the empirical interpretations of any particular axiomatic system of physics. The thesis was completed in 1955 and published in 1959.

2. UTILITY THEORY AND GAME THEORY

Although Adams' dissertation was on the foundations of physics, his first publication was a paper on game theory written jointly with Duncan Luce (1956). He had some work many years ago that he didn't publish that I found extremely interesting on the game of nim, and his interest in utility theory and game theory has continued over a long period. There was an important paper on riskless choice written with Bob Fagot in 1959. In

1960 he published a detailed survey of Bernoullian utility theory. There was also an early paper in 1961, in two parts, on rational betting systems in the more general Bayesian setting. A good many years later, in 1980, he published a detailed application of Dick Jeffrey's decision model to rational betting and information acquisition. That paper was published jointly with Roger Rosenkrantz.

It is obvious that Adams' interest in and knowledge about the literature on utility theory and subjective probability extend over many years and several contributions. Moreover, a number of the papers to be discussed under other headings bear on problems of measuring utility or subjective probability.

3. General Measurement Theory

As in the case of many of us, Adams' work on utility theory was very much intertwined with work on measurement theory. There is, however, an important special flavor to his work on measurement theory to be found already in the first article in 1965, written with Bob Fagot and Dick Robinson, on a theory of appropriate statistics. This same theme is to be found in a 1970 article written jointly with the same two authors entitled "On the Empirical Status of Axioms in Theories of Fundamental Measurement." These two papers made an important contribution to detailed discussions of the concept of meaningfulness in the theory of measurement, a topic of continued and controversial interest. At the same time, Adams published his own skeptical views of the representational theory of measurement in a 1965 article, "Elements of a Theory of Inexact Measurement," and in a 1966 article, "On the Nature and Purpose of Measurement." He has played the role of a skeptical critic, but a constructive one, for much of the standard work in the theory of measurement over many years. This line of work is represented in a paper in 1974, given at the Tarski symposium, another paper with Bob Fagot in 1975 on biassed bisection operations, a 1979 general paper on measurement theory, and a 1979 paper with I. F. Carlstrom on representing approximate ordering and equivalence relations.

These many articles amply demonstrate that Adams is one of the few philosophers of his generation with a thorough technical knowledge of the literature on measurement and an original philosophical viewpoint about the developments in the subject. In all likelihood, this work is little known to many philosophers familiar with his work on the logic of conditionals.

Finally, I should mention a recent surprising book, written with his brother William Adams, entitled *Archeological Typology and Practical*

Reality (1992). The problems of classification in archeology, as both Ernie and his brother have persuaded me, are subtle and complex, as is the theory of classification applied to almost any developed empirical subject. This work is characteristic of Adams' other work, in the sense of not being content with generalities, for it features thorough pursuit of important and critical details.

4. FOUNDATIONS OF GEOMETRY

Although he has not published as much on this topic, it is evident from conversations with Adams that his interest in the foundations of geometry is as serious and as deep as for any subject in which he has been involved. Furthermore, this work began early, with an article on the empirical foundations of elementary geometry in 1961. That article takes up a theme already begun in his dissertation: the problem of giving empirical interpretations of physical or geometrical axioms. It was some years before he published another article on geometry, but all during that time he was continuing to think deeply about the subject, and he published one of the ideas for which he is best known by those involved in the foundations of geometry, namely, how topological ideas can be connected to naive conceptions of surface. The article appeared in 1973. In my judgment, it is one of the most interesting things written on this fundamental perceptual problem. The only recent thing he has published about geometry is a review of Avrum Stroll's book *Surfaces* (1989), but he has a substantial body of work in preparation.

In my opinion, the unpublished manuscripts I have read constitute the most significant body of work on the foundations of geometry by any philosopher in the past several decades. What is especially important about this work is the detailed development of a theme that runs through Adams' work, namely, how to make actual practice rigorous – whether it be a matter of using conditionals in speech, or using geometrical concepts in ordinary talk.

This keen eye for the detailed analysis of the discrepancy between too simple theories and complex practice is a dominant aspect of his work on geometry. I mention especially his recent unpublished work on superposition and his continued emphasis over many years on the epistemic priority of topology. His earlier ideas, critical of general theories of measurement, can also be seen at work in his remarks on classical theories of geometry, from Hilbert to Tarski, insofar as they are systems to be applied to the world.

There is a lot more that I would like to say about his geometry, but this is not really the appropriate occasion. I do personally look forward to the appearance of several substantial manuscripts that are just about to surface.

5. LANGUAGE AND LOGIC, ESPECIALLY CONDITIONALS

Adams' work on conditionals is well known and does not need a general review by me. I do want to note that the work began early, with the appearance of the article "On the Reasonableness of the Inferences Involving Conditionals" in 1964. Although the probabilistic interpretation of conditionals is not unique with him, it is certainly the case that he has been one of the most prominent exponents, and he has exploited well his thorough knowledge of the work over the same period in the foundations of probability to make important connections.

NOTE

The references to Adams' papers here are their dates of publication. The complete list of his publications is given at the end of this volume.

Publications of Ernest W. Adams

(1956) "The Determination of the Subjective Characteristic Functions of Games with Misperceived Payoff Functions," with R. D. Luce. *Econometrica* 24:158–71.

(1958) "An Axiomatic Formulation and Generalization of Successive Intervals Scaling," with S. Messick. *Psychometrica* 23:3–16.

(1959) "A Model of Riskless Choice," with R. F. Fagot. *Behavioral Science* 4:1–10.

(1959) "The Foundations of Rigid Body Mechanics and the Derivation of Its Laws from Those of Particle Mechanics," in L. Henkin, P. Suppes & A. Tarski (eds.), *Symposium on the Axiomatic Method*, pp. 250–65. North Holland, Amsterdam.

(1960) "A Survey of Bernoullian Utility," in H. Soloman (ed.), *Mathematical Thinking in the Measurement of Behavior*, pp. 151–268. Free Press, New York.

(1961) "The Empirical Foundations of Elementary Geometry," in H. Feigl & G. Maxwell (eds.), *Current Issues in the Philosophy of Science*, pp. 197–226. Holt, Rinehart & Winston, New York.

(1961) "On Rational Betting Systems." *Archiv für Mathematische Logik und Grundlagensforschung*, 6:112–28.

(1961) "On the Reasonableness of Inferences Involving Conditionals," in F. Larroyo & J. L. Curiel (eds.), *Memorias des XIII Congresso International de Filosofia*, vol. 5, pp. 1–10. Cummunicaciones Libres, Universidad Nacional Autonoma do Mexico.

(1965) "A Theory of Appropriate Statistics," with R. F. Fagot & R. E. Robinson. *Psychometrika* 39:99–127.

(1965) "The Logic of Conditionals," *Inquiry* 8:166–97.

(1965) "Elements of a Theory of Inexact Measurement." *Philosophy of Science* 32:205–28.

(1966) "On the Nature and Purpose of Measurement." *Synthese* 16:225–69.

(1966) "Probability and the Logic of Conditionals," in J. Hintikka & P. Suppes (eds.), *Aspects of Inductive Logic*, pp. 265–316. North Holland, Amsterdam.

(1970) "On the Empirical Status of Axioms in Theories of Fundamental Measurement," with R. F. Fagot & R. E. Robinson. *Journal of Mathematical Psychology*, 7:379–409.

(1970) "Subjective and Indicative Conditionals." *Foundations of Language* 6:89–94.

(1973) "The Naive Conception of the Topology of the Surface of a Body," in P. Suppes (ed.), *Space, Time and Geometry*, pp. 402–24. D. Reidel, Dordrecht.

(1974) "The Logic of 'Almost All," *Journal of Philosophical Logic* 3:3–17.

(1974) "Model-Theoretic Aspects of Fundamental Measurement Theory," in L. Henkin et al. (eds.), *Proceedings of the Tarski Symposium*, pp. 437–46. American Mathematical Society Publications, vol. 25. AMS, Providence, R.I.

(1975) "On the Uncertainties Transmitted from Premises to Conclusions in Deductive Inferences," with H. Levine. *Synthese* 30:429–60.

(1975) *The Logic of Conditionals*. D. Reidel, Dordrecht.

(1975) "A Theory of Biassed Bisection Operations and Their Inverses," with R. F. Fagot. *Journal of Mathematical Psychology* 12: 35–52.

(1976) "Prior Probabilities and Counterfactual Conditionals," in W. Harper & C. Hooker (eds.), *Foundations of Probability Theory, Statistical Inference, and Statistical Theories of Science*, vol. 1, pp. 1–21. D. Reidel, Dordrecht.

(1977) "A Note on Comparing Probabilistic and Modal Logics of Conditionals." *Theoria* 43: 186–94.

(1978) "Two Aspects of Physical Identity." *Philosophical Studies* 34: 111–34.

(1979) "Measurement Theory," in P. D. Asquith & H. E. Kyburg (eds.), *Current Research in Philosophy of Science*, pp. 207–27. Philosphy of Science Association, East Lansing, Mich.

(1979) "Representing Approximate Ordering and Equivalence Relations," with I. F. Carlstrom. *Journal of Mathematical Psychology* 19: 182–207.

(1980) "Applying the Jeffrey Decision Model to Rational Betting and Information Acquisition," with R. Rosenkrantz. *Theory and Decision* 12: 1–20.

(1981) "Truth, Proof, and Conditionals." *Pacific Philosophical Quarterly* 62: 323–39.

(1981) "Transmissible Improbabilities and Marginal Essentialness of Premises in Inferences Involving Indicative Conditionals." *Journal of Philosophical Logic* 10: 149–77.

(1982) "Approximate Generalizations and Their Idealization," in P. D. Asquith & T. Nickles (eds.), *PSA 1982*, vol. 1, pp. 199–207. Philosophy of Science Association, East Lansing, Mich.

(1983) "Probabilistic Enthymemes." *Journal of Pragmatics* 7: 283–95.

(1984) "Remarks on Convention T's Pragmatic and Semantic Associations, and Some of Its Limitations." *Pacific Philosophical Quarterly* 65: 124–39.

(1984) "On the Superficial." *Pacific Philosophical Quarterly* 65: 386–407.

(1986) "On the Dimensionality of Surfaces, Solids and Spaces." *Erkenntnis* 24: 137–201.

(1986) "Continuity and Idealizability of Approximate Generalizations." *Synthese* 67: 439–76.

(1986) "On the Logic of High Probability." *Journal of Philosophical Logic* 15: 255–79.

(1986) "Remarks on the Semantics and Pragmatics of Conditionals," in E. Traugott et al. (eds.), *On Conditionals*, pp. 169–78. Cambridge University Press.

(1986) "Problems of the Theory of Approximate Representation," in *Proceedings of the Sixteenth Annual Pittsburgh Conference on Modelling and Simulation*, vol. 16, pp. 541–6. Instrument Society of America, Pittsburgh.

(1987) "On the Meaning of the Conditional." *Philosophical Topics* 15: 5–22.

(1987) "Monotonicity, Convexity and Other Qualitative Psychophysical Laws," with R. F. Fagot. *Journal of Mathematical Psychology* 31: 113–34.

(1987) "Purpose and Scientific Concept Formation," with W. Y. Adams. *British Journal for Philosophy of Science* 38: 419–40.

(1988) "*Modus Tollens* Revisited." *Analysis (New Series)* 38: 122–8.

(1988) "Consistency and Decision: Variations on Ramseyan Themes," in W. L. Harper & B. Skyrms (eds.), *Causation in Decision, Belief Change, and Statistics, II*, pp. 49–69. Kluwer, Dordrecht.

(1988) "Confirming Inexact Generalizations," in A. Fine & J. Leplin (eds.), *PSA 1988*, Vol. I, pp. 10–16. Edwards Brothers, Ann Arbor, Mich.

(1988) "A Note on Solidity." *Australasian Journal of Philosophy* 66: 512–16.

(1991) *Archeological Typology and Practical Reality*, with William Y. Adams (senior author). Cambridge University Press.

(1992) "On the Empirical Status of Measurement Axioms: the Case of Subjective Probability," in C. Wade Savage & P. Ehrlich (eds.), *The Nature and Function of Measurement*, pp. 53–74. Erlbaum, Hillsdale, N.J.

(1992) "Grice on Indicative Conditionals." *Pacific Philosophical Quarterly* 73: 1–15.

Papers to Appear

"Classical Physical Abstraction," to appear in *Erkenntnis*.

"Practical Causal Generalizations," to appear in P. Humphreys (ed.), *Patrick Suppes: Scientific Philosopher*.

"Positive, Comparative, and Superlative," to appear in *The Notre Dame Journal of Formal Logic*.

Survey article "Conditionals," with Ruth Manor, to appear in D. Gabbay et al. (eds.), *Handbook of Philosophy of Language*.

"Conditional Information," to appear in D. Dubois, I. R. Goodman, & P Calabrese (eds.), special issue of *IEEE Transactions on Systems, Man and Cybernetics* on *Conditional Event Algebra/Conditional Probability Logic*.

Reviews

H. Simon, *Models of Man*, review in *Journal of Philosophy* (ca. 1960); E. Nagel, P. Suppes, & A. Tarski (eds.), *Logic, Methodology and Philosophy of Science*, review in *Journal of Philosophy* (1962); J. Hintikka & P. Suppes (eds.), *Information and Inference*, review in *Synthese* (ca. 1965); A. Grunbaum, *Geometry and Chronometry in Philosophical Perspective*, review in *Philosophy of Science* (1968); F. Jackson, *Conditionals*, review in *Philosophical Review* (July 1990); Ilke Niiniluoto, *Truthlikeness*, review in *Synthese* (1990); R. Jeffrey, *Probability and the Art of Judgement*, review in *Journal of Philosophy* (March 1993).